Stem Cell Biology and Regenerative Medicine

Series Editor
Kursad Turksen, Ph.D.
kursadturksen@gmail.com

More information about this series at http://www.springer.com/series/7896

Rosalinda Madonna
Editor

Stem Cells and Cardiac Regeneration

Springer

Editor
Rosalinda Madonna
Institute of Cardiology
Department of Neurosciences and Imaging Laboratory
of Experimental Cardiology
Center of Excellence on Aging—CESI
"G. d'Annunzio" University
Chieti Scalo, Italy

ISSN 2196-8985 ISSN 2196-8993 (electronic)
Stem Cell Biology and Regenerative Medicine
ISBN 978-3-319-25425-8 ISBN 978-3-319-25427-2 (eBook)
DOI 10.1007/978-3-319-25427-2

Library of Congress Control Number: 2015958809

Springer Cham Heidelberg New York Dordrecht London

Printed on acid-free paper

Springer International Publishing AG Switzerland is part of Springer Science+Business Media (www.springer.com)

To our families and to our coworkers—our extended families

Preface

Ischemic heart disease (IHD) and heart failure (HF) are the major causes of morbidity and mortality in Europe and worldwide. IHD and HF patients could benefit from therapies that would accelerate natural processes of postnatal collateral vessel formation and/or muscle regeneration. Although treatment for acute myocardial infarction has improved over the past decades, there are currently no effective therapies for the associated problem of IHD and HF. Therefore, novel therapeutic strategies are required in order to reduce the extent of myocardial infarction, preserve cardiac function, and improve clinical outcomes in patients with IHD. The purpose of this book is to provide an update of recent advances in the basic and clinical applications of cell-based therapies for myocardial repair and regeneration in IHD and HF. In this book, cardiologist clinicians and scientists will have an opportunity to learn about the potential of multipotent and pluripotent stem cells and all stem cell-based regenerative therapies, aimed at treating these diseases. The first and second parts of the book cover the basic aspects of stem cells, while subsequent sections will be more dedicated to the clinical application of stem cells for the treatment of IHD and HF. In particular, Chaps. 1–4 provide definitions, isolation criteria, and characterization of embryonic and adult stem cells, as well as pluripotent stem cells and tissue-specific progenitor and stem cells (including adipose, bone, lung, brain, blood, heart tissues). Chapters 5–9 examine the role of critical regulators of stem cell differentiation in myocardial regeneration that include circadian rhythms, microRNAs, epigenetics, microvesicles, and exosomes. Chapter 10 assesses the influence of cell delivery routes and injectable biomaterials on the engraftment and efficacy of transplanted stem cells. Finally, Chap. 11 critically reviews the use of stem/progenitor cells in cardiac regeneration and discusses current controversies, unresolved issues, challenges, and future directions.

The book is the result of collaborative efforts from all members of the European Society of Cardiology Working Group on Cellular Biology of the Heart and different investigators in the stem cell research field around the world. I would like to thank all authors for their contributions.

Chieti Scalo, Italy Rosalinda Madonna, M.D., Ph.D.

Acknowledgments

This book would not have been possible without the collaborative efforts, abnegation, and persistence from all members of the European Society of Cardiology (ESC) Working Group (WG) on Cellular Biology of the Heart, on which behalf the book is published. The ESC-WG is deeply grateful to external authors (Silvia Agostini, Valentina Casieri, Neil Davies, Christina F. de Veij Mestdagh, Pieterjan Dierickx, Federica Diofano, Pieter A. Doevendans, Martijn J. H. Doeleman, Bastiaan Du Pré, Dries A.M. Feyen, Thomas Franz, Niels Geijsen, Kyle Goetsch, Vincenzo Lionetti, Marco Matteucci, Francesco Moccia, Malebogo Ngoepe, Gaia Papini, Paola Rebuzzini, Toon van Veen, Estella Zuccolo) for their expert contributions.

Contents

Contributors

Silvia Agostini Laboratory of Medical Science, Institute of Life Sciences, Scuola Superiore Sant'Anna, Pisa, Italy

Valentina Casieri Laboratory of Medical Science, Institute of Life Sciences, Scuola Superiore Sant'Anna, Pisa, Italy

Sean Davidson The Hatter Cardiovascular Institute, London, UK

Neil Davies Cardiovascular Research Unit, Department of Surgery, Faculty of Health Sciences, University of Cape Town, Observatory, South Africa

Christina F. de Veij Mestdagh Division Heart and Lung, Cardiology, Experimental Cardiology Laboratory, University Medical Center Utrecht, Utrecht, The Netherlands

ICIN, Netherlands Heart Institute, Utrecht, The Netherlands

Pieterjan Dierickx Division of Heart and Lungs, Department of Cardiology, University Medical Center Utrecht, Utrecht, The Netherlands

Hubrecht Institute-KNAW and University Medical Center, Utrecht, The Netherlands

Federica Diofano Laboratory of General Physiology, Department of Biology and Biotechnology "Lazzaro Spallanzani", University of Pavia, Pavia, Italy

Pieter A. Doevendans Division of Heart and Lungs, Department of Cardiology, University Medical Center Utrecht, Utrecht, The Netherlands

Martijn J.H. Doeleman Division Heart and Lung, Cardiology, Experimental Cardiology Laboratory, University Medical Center Utrecht, Utrecht, The Netherlands

Bastiaan Du Pré Hubrecht Institute-KNAW and University Medical Center, Utrecht, The Netherlands

Division of Heart and Lungs, Department of Medical Physiology, University Medical Center Utrecht, Utrecht, The Netherlands

Felix B. Engel Department of Nephropathology, Laboratory of Experimental Renal and Cardiovascular Research, Institute of Pathology, Friedrich-Alexander-Universität Erlangen-Nürnberg (FAU), Erlangen, Germany

Peter Ferdinandy Department of Pharmacology and Pharmacotherapy, Semmelweis University, Budapest, Hungary

Pharmahungary Group, Szeged, Hungary

Dries A.M. Feyen Division Heart and Lung, Cardiology, Experimental Cardiology Laboratory, University Medical Center Utrecht, Utrecht, The Netherlands

Thomas Franz Division of Biomedical Engineering, Department of Human Biology, Faculty of Health Sciences, University of Cape Town, Observatory, South Africa

Niels Geijsen Hubrecht Institute-KNAW and University Medical Center, Utrecht, The Netherlands

Faculty of Veterinary Medicine, Clinical Sciences of Companion Animals, Utrecht University, Utrecht, The Netherlands

Kyle Goetsch Cardiovascular Research Unit, Department of Surgery, Faculty of Health Sciences, University of Cape Town, Observatory, South Africa

Sandrine Lecour Cardioprotection Group, Hatter Institute for Cardiovascular Research in Africa and MRC Inter-University Cape Heart Group, Faculty of Health Sciences, University of Cape Town, Observatory, South Africa

Vincenzo Lionetti Laboratory of Medical Science, Institute of Life Sciences, Scuola Superiore Sant'Anna, Pisa, Italy

Rosalinda Madonna Institute of Cardiology, Department of Neurosciences and Imaging Laboratory of Experimental Cardiology, Center of Excellence on Aging—CESI, "G. d'Annunzio" University, Chieti Scalo, Italy

Marco Matteucci Laboratory of Medical Science, Institute of Life Sciences, Scuola Superiore Sant'Anna, Pisa, Italy

Francesco Moccia Laboratory of General Physiology, Department of Biology and Biotechnology "Lazzaro Spallanzani", University of Pavia, Pavia, Italy

Malebogo Ngoepe Department of Mechanical Engineering, University of Cape Town, Rondebosch, South Africa

Gaia Papini Laboratory of Medical Science, Institute of Life Sciences, Scuola Superiore Sant'Anna, Pisa, Italy

Paola Rebuzzini Laboratory of Developmental Biology, Department of Biology and Biotechnology "Lazzaro Spallanzani", University of Pavia, Pavia, Italy

Rainer Schulz Institute of Physiology, Justus-Liebig Giessen University of Giessen, Giessen, Germany

Joost P.G. Sluijter Division Heart and Lung, Cardiology, Experimental Cardiology Laboratory, University Medical Center Utrecht, Utrecht, The Netherlands

ICIN, Netherlands Heart Institute, Utrecht, The Netherlands

Linda W. Van Laake Division of Heart and Lungs, Department of Cardiology, University Medical Center Utrecht, Utrecht, The Netherlands

Hubrecht Institute-KNAW and University Medical Center, Utrecht, The Netherlands

Toon van Veen Division of Heart and Lungs, Department of Medical Physiology, University Medical Center Utrecht, Utrecht, The Netherlands

Estella Zuccolo Laboratory of General Physiology, Department of Biology and Biotechnology "Lazzaro Spallanzani", University of Pavia, Pavia, Italy

Abbreviations

AGO1-4	Argonaute proteins
ALCAM	Activated leukocyte cell adhesion molecule
AMI	Acute myocardial infarction
ANF	Atrial natriuretic factor
Ang-1	Angiopoietin-1
bFGF	Basic fibroblast growth factor
BMP	Bone morphogenetic protein
Bmp1	Bone morphogenetic protein 1
BMSCs	Bone marrow stromal cells
CABG	Coronary artery bypass grafting
CAM	Chorioallantoic membrane
CBT	Cell based therapy
CDKs	Cyclin dependent kinases
CIRP	Cold inducible RNA-binding protein
CLOCK	Circadian Locomotor Output Cycles Kaput
CM	Cardiomyocytes
CPCs	Cardiac progenitor cells
CSCs	Cardiac stem cells
Dll4	Notch ligand Delta-like 4
DLS	Dynamic light scattering
DSB	Double-strand break
EBs	Embryoid bodies
EGF	Epidermal growth factor
eIFs	Initiation factors
EPCs	Circulating endothelial progenitor cells
EpiSCs	Epiblast stem cells
ES cells	Embryonic stem cells
ESRRB	Estrogen-related receptor beta
ET-1	Endothelin-1
EVs	Extracellular vesicles
FD	Familial Dysautonomia

FGF-2	Fibroblast growth factor 2
FZD7	Frizzled family receptor 7
GCSF	Granulocyte colony stimulating factor
HASF	Akt-induced stem cell factor
HAT	Histone acetyl transferase
HDAC	Histone deacetylase
HF	Heart failure
HGF	Hepatocyte growth factor
HLA	Human leukocyte antigen
iCMs	Induced cardiomyocyte-like cells
IGF-1	Insulin growth factor 1
IKAP	IkB kinase complex associated protein
Il-6	Interleukin-6
iPS cells	Induced pluripotent stem cells
IRAK	Interleukin-1 receptor-associated kinase
JAM-A	Junctional adhesion molecule-A
KLF-4	Kruppel-like factor-4
LIF	Leukemia inhibitory factor
LINE-1	Interspersed Nucleotide Elements-1
LncRNAs	Long non-coding RNAs
LQTS	Congenital long-QT syndrome
LVEF	Left ventricular ejection fraction
MAPK	Mitogen-activated protein kinase
Mef2c	Myocyte-specific enhancer factor 2c
MET	Mesenchymal-to-epithelial transition
α-MHC	α-Myosin heavy chain
β-MHC	β-Myosin heavy chain
miRNAs	MicroRNAs
MLC	Myosin light chain
MMPs	Matrix metalloproteinases
MSCs	Mesenchymal stem cells
MYDGF	Myeloid-derived growth factor
MYOCD	Myocardin
MyoD	Myoblast determination protein
NHEJ	Non-homologous end joining
NR5A-2	Nuclear receptor subfamily 5 group A member-2
NRG-1	Neuregulin-1
NTA	Nanoparticle tracking analysis
OCT-4	Octamer-binding transcription factor-4
PDGFRB	Platelet-derived growth factor receptor
PEG	Polyethylene glycol
PET	Positron emission tomography
PGE2	Prostaglandin E2
PLGF	Placental growth factor
PNIPAAm	Poly (*N*-isopropylacrylamide)

rAAV	Recombinant adeno-associated virus
RFP	Red fluorescent protein
RIC	Remote ischaemic preconditioning
RNAi	RNA interference
RORE	Retinoic acid-related orphan receptor response element
SAP	Self-assembling peptides
SCID	Severe combined immunodeficiency SCID mice
Sfrp2	Secreted frizzled related protein 2
SIRPA	Signal-regulatory protein alpha
SIRT-1	Sirtuin 1
SMA	Spinal muscular atrophy
SMN	Survival of Motor Neuron
SPECT	Single-photon emission computed tomography
SRA1	RNA activator 1
SRF	Serum response factor
SRM	Stem cell released molecules
SRY	Sex determining region Y-box 2
SSEA	Specific membrane antigens
TERT	Telomerase reverse transcriptase
TGF-β	Transforming growth factor β
TIPeR	Transcriptome-induced phenotype remodeling
TRBP	Transactivating response RNA-binding protein
VEGF	Vascular endothelial growth factor
VPA	Valproic acid
ZFN	Zinc finger nuclease

About the Author

Rosalinda Madonna is an Assistant Professor at Center of Excellence on Aging, G. d'Annunzio University in Chieti, Italy, and an Adjunct Assistant Professor in Cardiology, Department of Internal Medicine, University of Texas Medical School in Houston, Texas. She received both her MD in Medicine and Surgery and PhD in Biotechnologies and Biosciences and Cardiology Specialization at G. D'Annunzio University in Chieti, Italy. She worked as a Research Fellow at the University of Texas Medical School in Houston and then as a Research Scientist at Texas Heart Institute Center for Cardiovascular Biology and Atherosclerosis Research in Houston, Texas, and at the Institute of Molecular Cardiology, University of Louisville in Louisville, Kentucky. Author of over 90 papers and 15 book chapters, she is an ex officio nucleus member of the European Society of Cardiology Working Group in Cellular Biology of the Heart and Councilor of the Italian Society of Cardiovascular Research (SIRC). Her research interests include atherosclerosis, stem cells, and regenerative medicine.

Part I
Stem Cells Sources for Cardiac Regeneration. Basic Principles

Chapter 1
Biology and Function of Stem Cells

Rosalinda Madonna

1.1 Basics of Stem Cell Biology

Stem cells are a population of immature tissue precursor cells capable of self-renewal or proliferation as well as differentiation into a spectrum of different cell types under appropriate conditions. In general, they share the following characteristics: (1) a high capacity for self-renewal; (2) the potential for multipotent differentiation potential; (3) the ability to be cultured ex vivo and used for tissue engineering (reprogramming); and (4) plasticity (transdifferentiating ability) (Perin et al. 2003; Vats et al. 2005). The classification of stem cells is still evolving. On the basis of differentiating potential, stem cells can currently be classified into four categories: (1) totipotent, (2) pluripopotent, (3) multipotent, and (4) monopotent or oligopotent. Totipotent stem cells in general have the potential to differentiate into cells of all three main embryonal layers (ectodermal, endodermal, and mesodermal). One totipotent cell can differentiate into an intact organism with central and peripheral nervous systems if that cell is implanted into a functional uterus. In mammals, only zygotes and blastomeres at the early stage of cleavage are totipotent stem cells. With the progression of differentiation, zygotes (totipotent stem cells) form themselves into outer and inner layers of cells. The cells in the inner layer can give rise to every cell type in the body, but they cannot do so without the outer layer, which will become the placenta. The inner cells are called pluripotent stem cells. As pluripotent stem cells continue to divide, they begin to specialize further and become the progenitors of special tissues. At this stage, they are multipotent; in other words, they can differentiate into several types of cells within a given organ. For example, multipotent blood stem cell or hematopoietic stem cells can develop into red blood cells, white blood cells, or

R. Madonna (✉)
Institute of Cardiology, Department of Neurosciences and Imaging Laboratory of Experimental Cardiology, Center of Excellence on Aging—CESI, "G. d'Annunzio" University, Chieti Scalo, Italy
e-mail: rmadonna@unich.it

R. Madonna (ed.), *Stem Cells and Cardiac Regeneration*, Stem Cell Biology and Regenerative Medicine, DOI 10.1007/978-3-319-25427-2_1

platelets. Monopotent, or oligopotent, stem cells can only give rise to one or few types of specialized cells. For example, mesenchymal cells can differentiate into bone, muscle, fat, and some other connective tissues (Morrison et al. 1997; Tsai et al. 2002; Hsu et al. 2004). On the basis of their origins and biological properties, stem cells can be classified as either (1) embryonic (ES) stem cells or (2) adult stem cells. ES cells are derived from the inner layer mass of the blastocyst. Having potent differentiating ability, one single ES cell can develop into more than 200 kinds of cells, then into different tissues and organs. Its use in the medical arena has been a hot topic in recent years. ES cells possess the ability to proliferate in an undifferentiated state for prolonged periods in culture and the ability to differentiate into every tissue type under favorable conditions (Maltsev et al. 1994; Dinsmore et al. 1996; Chandross and Mezey 2002) thus, they are being considered as a new, promising source of donor cells for organ transplantation.

1.2 Pluripotent Versus Multipotent Stem Cells

ES cells can be harvested from three sources: aborted fetuses (cadaveric stem cells), embryos left over from in vitro fertilization (discarded embryos), and embryos created in the laboratory solely for the purpose of producing stem cells (research embryos). As animal studies have proved that ES cells can differentiate into cardiomyocytes in vitro, the use of human ES cell line as a novel source of human cardiomyocytes may become an attractive option. In vitro differentiation of human ES cells into cardiomyocytes has been demonstrated by Kehat et al. (2001). However, ethical issues have been raised against harvesting human ES cells, especially if this process requires destruction of an embryo that is considered as a potential human life by some ethical groups. However, the opposition to the use of human ES cells has been countered by arguments that their use will have profound therapeutic implications. Other potential obstacles to using ES cells are that recipients often need to receive immunosuppressants because ES cells are potentially allogenic. Uncontrolled differentiation of ES cells may cause other problems, such as the development of cardiac or vascular neoplasm. Transplanted ES cells may form teratomas if some undifferentiated totipotent cells are still present. Thomson et al. (1998) observed the formation of teratomas—i.e., tumors containing a mix of differentiated human cell types, including cells characteristic of ectoderm, mesoderm, and endoderm—in severe combined immunodeficiency (SCID) mice after injection with human ES cells. At present, the transfer of ES cell-derived cardiomyocytes remains a relatively new technique with limited experience so far. Adult stem cells are the undifferentiated cells that exist in a differentiated tissue or organ and that are capable of specializing into cells of the tissue or organ from which they originated. Their capacity for self-renewal allows tissues and organs to maintain stability of function (Jiang et al. 2002). Sources of adult stem cells include not only the regenerating tissues, such as heart, adipose tissue, bone marrow, blood, liver and epidermis, but also the non-divisive tissues, such as brain. Compared with ES cells, autologous adult stem cells are not faced with any major ethical or immunological controversies surrounding their use in the same individual from whom the stem cells were obtained. But their ability to proliferate and differentiate is less powerful than that of ES cells; they are often difficult to identify, isolate, and purify; and they are not numerous enough for use in transplantation without being expanded in vitro substantially. For example, there is only one hematopoietic cell for every 1000–5000 bone marrow stromal cells (BMSCs). Finally, adult stem cells do not replicate indefinitely in culture. Stem cells are defined by their capacity for continuous self-renewal (proliferation) and their ability to give rise to functionally or morphologically specialized somatic cells (differentiation). Stem cells exist abundantly in embryos, and their numbers decline dramatically after birth (Vats et al. 2005). However, in certain fast growing and renewable tissues, there are considerable numbers of stem cells that can be mobilized when tissues need to grow, regenerate, or repair themselves. As stated above, in general, stem cells in embryonic tissue have a higher potential for growth and differentiation than do those in adult tissue. Theoretically, ES cells may generate any type of cell found in adult tissues. The differentiation capacity is thus referred to as totipotency or pluripotency. Adult stem cells have a lower capacity for differentiation, and they usually produce only limited numbers of cell lineages; they are referred to as multipotent or oligopotent stem cells. In certain tissues, stem cells only give rise to a single type of somatic cells (Slack 2000). These stem cells are monopotent. The differentiation process is complex and often requires many steps. Therefore, there are many intermediate types of partially differentiated stem cells that are committed to certain cell lineages. The partially differentiated cells are precursor or progenitor cells. For instance, circulating endothelial progenitor cells (EPCs) can promptly differentiate into mature endothelial cells that replace dead or dying cells when the endothelium of arteries is injured by atherosclerosis (Urbich and Dimmeler 2004).

1.3 Basic Requirements of Stem Cell Sources for Cardiac Regeneration

Transplantation of stem cells has been used recently to repair injured myocardium, regenerate new myocytes and enhance vascularity. In addition to generation of neomyocardium, grafted stem cells can participate in remodeling and revascularization in the heart damaged by ischemic insults. This therapy represents a new frontier in the treatment of cardiovascular disease (Perin et al. 2003). Selection of a suitable type of stem cells will be key to the success of stem cell therapy. Usually, to achieve the most effective tissue repair or regeneration, cells used for transplantation should have the following characteristics: (1) high rates of survival and proliferation (i.e., they should be able to reach the injured area, stay alive, and proliferate in the injured tissues); (2) strong potency of differentiation (i.e., the cells could differentiate into a certain mature type to help repair the heart); and (3) potential for highly effective engraftment and integration with native or host cardiac cells (i.e., the cells should be able to contract after differentiation and form stable intercellular gap junctions and electrophysiological couplings with surrounding cardiac cells). Following differentiation, myocytes newly formed from the implanted stem cells should have myotubes whose electromechanical properties are the same as those of myotubes of native cardiomyocytes (Deb et al. 2003). Their contraction should remain synchronized during systole and diastole. In essence, the cell transplantation should achieve the goal of replacing damaged cardiac myocytes and, to a large extent, restore cardiac function. Currently, both embryonic and adult stem cells are used in experimental cardiac cell transplantation studies, while only adult stem cells (e.g., bone marrow-derived mesenchymal cells, skeletal myoblasts, endothelial progenitor cells) and fetal cardiomyocytes are used in clinical trials. Each stem cell type has unique biological properties that offer both advantages and limitations to use (Forrester et al. 2003). Therefore, selection of the most suitable stem cells for use in heart failure patients is still a major focus of current research.

1.4 Adult Stem Cell Plasticity

With the progress of stem cell research, many studies have revealed the presence of stem cells in adult somatic tissues, including the myocardium. Investigators examined the plasticity of adult stem cells, i.e., the ability of stem cells from one tissue to generate the specialized cell type(s) of another tissue. Some researchers believe that the plasticity of adult stem cells is closely related to the microenvironment around them, the so-called stem cell niche. Cell factors in the stem cell niche give rise to differentiation and help to establish communication among cells. Given the environment, some adult stem cells are capable of being "genetically reprogrammed" to generate specialized cells that are characteristic of different tissues. Transplantation of bone marrow cells induced angiogenesis, and if the bone marrow cells were cultured in 5-azacytidine containing medium, they had ability to differentiate into cardiac-like muscle cells both in culture and in vivo in ventricular scar tissue and improve myocardial function Zhang et al. (Yeh et al. Ciculation 2004) injected human adult peripheral blood CD34+ cells through the tail veins of SCID mice with infarcted myocardium. Two months later, cardiomyocytes, endothelial cells, and smooth muscle cells bearing human leukocyte antigen (HLA) were identified in the infarct and periinfarct regions of the mouse hearts. Although there is no evidence to prove that adult stem cells can regenerate all cell types in the body, their apparent plasticity sheds new light on cell transplantation therapy. Recently, however, (Murry et al. Nature 2004) used both cardiomyocyte restricted and ubiquitously expressed reporter transgenes to track the fate of hematopoietic stem cells after transplantation into normal and injured adult mouse hearts. They were unable to show that bone marrow-derived stem cells transdifferentiated into cardiomyocytes in their system. And they observed that stem cell-engrafted hearts had no sign for hematopoietic stem cell-derived cardiomyocytes when compared with sham-engrafted hearts. This study indicated that hematopoietic stem cells may not readily generate a cardiac phenotype and raises a cautionary note for clinical studies of infarct repair.

References

Chandross K, Mezey E (2002) Plasticity of adult bone marrow stem cells. Adv Cell Aging Gerontol 9:73–95

Deb A, Wang S et al (2003) Bone marrow-derived cardiomyocytes are present in adult human heart: a study of gender-mismatched bone marrow transplantation patients. Circulation 107(9):1247–1249

Dinsmore J, Ratliff J et al (1996) Embryonic stem cells differentiated in vitro as a novel source of cells for transplantation. Cell Transplant 5(2):131–143

Forrester JS, Price MJ et al (2003) Stem cell repair of infarcted myocardium: an overview for clinicians. Circulation 108(9):1139–1145

Hsu C-P et al (2004) Cell therapy for cardiac diseases-bridging basic and clinical science. Acta Cardiol Sin 20:59–72

Jiang Y, Jahagirdar BN et al (2002) Pluripotency of mesenchymal stem cells derived from adult marrow. Nature 418(6893):41–49

Kehat I, Kenyagin-Karsenti D et al (2001) Human embryonic stem cells can differentiate into myocytes with structural and functional properties of cardiomyocytes. J Clin Invest 108(3):407–414

Maltsev VA, Wobus AM et al (1994) Cardiomyocytes differentiated in vitro from embryonic stem cells developmentally express cardiac-specific genes and ionic currents. Circ Res 75(2):233–244

Morrison SJ, Shah NM et al (1997) Regulatory mechanisms in stem cell biology. Cell 88(3):287–298

Murry CE, Soonpaa MH et al (2004) Haematopoietic stem cells do not transdifferentiate into cardiac myocytes in myocardial infarcts. Nature 428(6983):664–668

Perin EC, Geng YJ et al (2003) Adult stem cell therapy in perspective. Circulation 107(7):935–938

Slack JM (2000) Stem cells in epithelial tissues. Science 287(5457):1431–1433

Thomson JA, Itskovitz-Eldor J et al (1998) Embryonic stem cell lines derived from human blastocysts. Science 282(5391):1145–1147

Tsai RY, Kittappa R et al (2002) Plasticity, niches, and the use of stem cells. Dev Cell 2(6): 707–712

Urbich C, Dimmeler S (2004) Endothelial progenitor cells: characterization and role in vascular biology. Circ Res 95(4):343–353

Vats A, Bielby RC et al (2005) Stem cells. Lancet 366(9485):592–602

Zhang S, Wang D et al (2004) Both cell fusion and transdifferentiation account for the transformation of human peripheral blood CD34-positive cells into cardiomyocytes in vivo. Circulation 2004 110(25):13803–7

Chapter 2
Embryonic Stem Cells for Cardiac Regeneration

Francesco Moccia, Federica Diofano, Paola Rebuzzini, and Estella Zuccolo

2.1 Introduction

Cardiovascular disease (CD) is the leading cause of death and morbidity in developed countries. Heart failure (HF) is the common end point of virtually all cardiac disorders, including acute myocardial infarction (AMI), coronary artery disease, hypertension, stroke, and diabetes, and affects about 1–2 % of the population in the Western world. The incidence of HF dramatically increases in ageing subjects and, consequently, represents the most frequent cause of hospitalization for those aged 65 and over (Pfister et al. 2014). Unlike smooth and skeletal muscle cells, adult cardiac myocytes show a limited regenerative capacity and cannot restore ventricular function to supply sufficient blood flow to the body (Jakob and Landmesser 2013; Pfister et al. 2014; Moccia et al. 2014). Heart transplantation (HT) represents the ultimate approach to treat end-stage HF, but this therapy is invasive, expensive and not suitable for elderly patients affected by comorbidities. Moreover, donor organ shortage is a major limitation factor worldwide and causes the death of an increasing number of subjects with head-stage HF who are listed for HT. Cell based therapy (CBT) provides an alternative option to replace damaged myocardium and vasculature and restore cardiac contractility (Moccia et al. 2013, 2014; Jakob and Landmesser 2013; Pfister et al. 2014). Several cardiac and non-cardiac stem/precursor/progenitor cells have been used in both pre-clinical studies and clinical trials to regenerate the failing heart, including bone marrow-derived mononuclear cells (MNCs), skeletal myoblasts, mesenchymal stem cells (MSCs), endothelial progenitor cells (EPCs), and

F. Moccia (✉) • F. Diofano • E. Zuccolo
Laboratory of General Physiology, Department of Biology and Biotechnology "Lazzaro Spallanzani", University of Pavia, via Forlanini 6, Pavia 27100, Italy
e-mail: francesco.moccia@unipv.it

P. Rebuzzini
Laboratory of Developmental Biology, Department of Biology and Biotechnology "Lazzaro Spallanzani", University of Pavia, via Forlanini 6, Pavia, 27100, Italy

R. Madonna (ed.), *Stem Cells and Cardiac Regeneration*, Stem Cell Biology and Regenerative Medicine, DOI 10.1007/978-3-319-25427-2_2

cardiac progenitor cells (CPCs) (Moccia et al. 2013, 2014; Jakob and Landmesser 2013; Pfister et al. 2014). Unfortunately, there is no evidence that these cell populations, even when harvested from patients' heart, such as CPCs, engraft within damaged myocardium, acquire a contractile phenotype and behave as genuine myocytes. Moreover, most of these studies did not show any remarkable improvement of cardiac performance: they recorded a modest 5 % increase, if any, in left ventricular ejection fraction (LVEF) of recipient hearts, which was mainly due to the paracrine release of factors and inhibit apoptosis, stimulate local angiogenesis and extracellular matrix remodelling, and recruit endogenous CPCs (Parsons 2012; Menasché 2012). This led to a paradigmatic shift in the field, according to which CBT should rather be carried out by using cardiac lineage-committed stem cells that are physically capable of replacing lost native CMs and functionally integrating within the beating heart. This novel context has led to the re-appreciation of the therapeutic use of pluripotent embryonic stem cells (ESCs), unspecialized cells which can be indefinitely cultivated in vitro, maintaining their features (undifferentiated status and self-renewal) and their intrinsic ability to generate all the terminal differentiated cell types, including cardiomyocytes (CMs). Therefore, ESCs hold tremendous potential for treating human disorders featuring tissue loss and insufficiency, such as HF and AMI (Wong and Bernstein 2010; Parsons 2012; Menasché 2012). Accordingly, both murine and human ESCs (mESCs and hESCs, respectively) differentiate into contracting myocytes that can integrate within damaged myocardium when properly manipulated in vitro (Kehat et al. 2001, 2004; Boheler et al. 2002; Passier et al. 2008). The present chapter aims at addressing the basic issues regarding isolation, characterization and CM differentiation of mESCs and hESCs and at focusing on their advantages and disadvantages for cardiac regeneration (Tables 2.1 and 2.2).

2.2 Murine Embryonic Stem Cells Isolation, Characterization and CM Differentiation

Mouse ESCs are undifferentiated, self-renewing and undifferentiated cells, derived from the inner cell mass (ICM or epiblast) of a blastocyst (Evans and Kaufman 1981; Martin 1981). Blastocysts, obtained or by natural (4 days *post coitum*) or by in vitro fertilization, are singly plated on plates coated with a feeder layer of immortalized or embryonic mouse fibroblasts and with an appropriate culture medium containing the leukemia inhibitory factor (LIF), necessary to repress mESC differentiation. To date, the presence of the feeder layer is necessary because of its ability to release the trophic factor, whereas LIF, a soluble glycoprotein belonging to the interleukin (IL)-6 family, is fundamental for mimicking the in vivo conditions in which it is released by the trophoectoderm cells and is perceived by the ICM cells, where it triggers the signal transduction finally activating the STAT3 transcription factor. STAT3 in turn transcribes all those genes involved in the maintenance of the undifferentiated state (Smith et al. 1988; Wobus and Boheler 2005; Graf et al. 2011).

After 2–4 days of culture, the rupture of the zona pellucida enables the outgrowth of both ICM cells and trophoblast cells, which are readily identifiable by phase-contrast

Table 2.1 Approaches to enhance cardiomyocyte yield from embryonic stem cells (ESCs)

Approach	Comments	References
Growth factors	Retinoic acid enhanced cardiomyocyte differentiation in mouse ESCs	Wobus et al. (1997)
	Exogenous glucose, amino acids, vitamins, and selenium enhanced cardiomyocyte differentiation in mouse ESCs	Guan et al. (1999)
	LIF enhances and inhibits cardiomyocyte commitment and proliferation in mouse ESCs in a developmental stage-dependent manner	Bader et al. (2000)
	Reactive oxygen species enhanced cardiomyocyte differentiation in mouse ESCs	Sauer et al. (2000), Buggisch et al. (2007), Sharifpanah et al. (2008), Wo et al. (2008)
	Endoderm enhanced cardiomyocyte differentiation in mouse ESCs	Bader et al. (2001), Rudy-Reil and Lough (2004)
	A TGF/BMP paracrine pathway enhanced cardiomyocyte differentiation in mouse ESCs	Behfar et al. (2002)
	Activation of the MEK/ERK pathway enhanced cardiomyocyte differentiation in mouse ESCs	Kim et al. (2007)
	Verapamil and cyclosporine enhanced cardiomyocyte differentiation in mouse ESCs	Sachinidis et al. (2006)
	5-Aza-2¢-deoxycytidine enhanced cardiomyocyte differentiation in human ESCs	Xu et al. (2002)
	Endoderm cell lines enhanced cardiomyocyte differentiation in human ESCs	Mummery et al. (2003, 2007)
	Ascorbic acid enhanced cardiomyocyte differentiation in human ESCs	Takahashi et al. (2003)
	Directed differentiation with activin A and BMP4 in monolayers of human ESC	Laflamme et al. (2007)
Genetic engineering	Lineage-restricted drug resistance gene resulted in highly purified cardiomyocyte cultures from mouse ESCs	Klug et al. (1996), Zandstra et al. (2003), Schroeder et al. (2005)
	Highly purified cardiomyocyte cultures generated by FACS of mouse ESCs expressing a lineage-restricted EGFP reporter	Müller et al. (2000)
	Targeted expression of a-1,3-fucosyltransferase enhanced cardiomyocyte differentiation in mouse ESCs	Sudou et al. (1997)
	Coexpression of EA1, dominant negative p53, and dominant negative CUL7 enhanced cell cycle in mouse ESC-derived cardiomyocytes	Pasumarthi et al. (2001)
	Expression of SV40 T antigen enhanced cell cycle in mouse ESC-derived cardiomyocytes	Huh et al. (2001)
	Antagonization of Wnt/b-catenin enhanced cardiomyocyte differentiation in mouse ESCs	Singh et al. (2007)
	Lineage-restricted drug resistance gene resulted in highly purified cardiomyocyte cultures from human ESCs	Anderson et al. (2007), Xu et al. (2008b)

(continued)

Table 2.1 (continued)

Approach	Comments	References
Miscellaneous	A single 90-s electrical pulse applied to day 4 EBs increased cardiomyocyte differentiation in mouse ESCs	Sauer et al. (1999)
	Application of mechanical loading enhanced cardiomyocyte differentiation in mouse ESCs	Shimko and Claycomb (2008), Gwak et al. (2008)
	FACS for transient Flk-1 isolated cardiomyogenic progenitors from mouse ESCs	Kattman et al. (2006)
	Cardiomyocyte enrichment using density centrifugation and cultures of cell aggregates in human ESCs	Xu et al. (2006)
	Activin A, BMP4, bFGF, VEGF, and DKK1 treatment, followed by KDR+	Yang et al. (2008)
	/c-kit–FACS, identified cardiovascular progenitor cells in human ESCs	

LIF leukemia inhibitory factor, *TGF* transforming growth factor, *BMP* bone morphogenetic protein, *MEK* mitogen-activated protein kinase, *ERK* extracellular-signal-regulated kinase, *FACS* fluorescence-activated cell sorting, *EGFP* enhanced green fluorescent protein, *EBs* embryoid bodies, *bFGF* basic fibroblast growth factor, *VEGF* vascular endothelial growth factor

microscopy. Clusters of ICM cells are thus mechanically isolated, dispersed and replated onto MEF feeder layers until clonal mESC line is established. In the presence of the feeder layer and LIF, mESCs may be cultured for a long time maintaining their unlimited proliferation capacity, pluripotency and ability to differentiate into all cell lineages (Hescheler et al. 1997; Boheler et al. 2002; Wobus and Boheler 2005). This is consistent with their ability to create teratomas comprising multiple cell types from the three germ layers upon injection into immunodeficient mice (Passier et al. 2006). It should, however, be pointed out that, unlike earlier predictions (Boheler et al. 2002), recent studies have demonstrated that (sub)chromosomal aberrations occur during mESC cultures (Neri et al. 2007; Rebuzzini et al. 2007; Gaztelumendi and Nogués 2014). Undifferentiated mESCs are characterized by a panel of cell surface and molecular markers that define their stemness; these include specific membrane antigens (SSEA-1,3) and membrane-bound receptors (gp130) and specific enzyme activities, such as telomerase and alkaline phosphatase (ALP). The transcriptional circuit responsible for self-renewal and pluripotency is governed by the epiblast/germ cell-restricted transcription factor Oct-3/4, whose expression levels must be maintained within an appropriate range to maintain stem cell self-renewal. A small elevation in Oct-3/4 levels (<2-fold) causes ESC differentiation into primitive endoderm or mesoderm, while suppressing Oct-3/4 expression induces them to acquire a trophectoderm phenotype. Oct-3/4 sustains self-renewal and pluripotency by acting in concert with Sox2, FoxD3, bone morphogenetic protein (BMP)-induced Smad-mediated activation of Id (inhibition of differentiation target genes), and LIF-dependent JAK2/Stat3 cascade. Additional molecular markers of pluripotency are represented by Nanog, Sox2 and Rex-1 that are expressed in the ICM, but are down-regulated during differentiation.

Table 2.2 Comparison of some properties of mouse and human embryonic stem cells

Marker	Mouse ES cells	Human ES cells	Reference
SSEA-1	+	–	Solter and Knowles (1978)
SSEA-3/-4	–	+	Thomson et al. (1998), Reubinoff et al. (2000), Xu et al. (2001), Henderson et al. (2002)
TRA-1-60/81	–	+	Thomson et al. (1998), Reubinoff et al. (2000), Xu et al. (2001), Henderson et al. (2002)
TRA-2-54	–	+	Henderson et al. (2002)
GCTM-2	–	+	Pera et al. (2000), Reubinoff et al. (2000)
TG 343	?	+	Henderson et al. (2002)
TG 30	?	+	Pera et al. (2000)
CD 9	+	+	Pera et al. (2000)
CD133/prominin	+	+	Kania et al. (2005), Carpenter et al. (2004)
Alkaline phosphatase	+	+	Wobus et al. (1984), Thomson et al. (1998)
Oct-4	+	+	Thomson et al. (1998), Pesce et al. (1999)
Nanog	+	+	Chambers et al. (2003), Mitsui et al. (2003)
Sox-2	+	+	Avilion et al. (2003), Ginis et al. (2004)
FGF4	+	–	Ginis et al. (2004)
LIF receptor	+	+/–	Richards et al. (2004)
Telomerase activity	+	+	Thomson et al. (1998), Armstrong et al. (2000)
Regulation of self-renewal	Via gp 130 receptors, MEF feeder layer, Nanog, BMP-4	Feeder cells (MEF or human cells), serum, bFGF, Matrigel	Niwa et al. (1998), Thomson et al. (1998), Xu et al. (2001), Chambers et al. (2003), Ying et al. (2003)
Growth characteristics in vitro	Tight, rounded, multilayer clusters	Flat, loose aggregates	Thomson et al. (1998)
EB formation	Simple and cystic EBs	Cystic EBs	Doetschman et al. (1985), Thomson et al. (1998), Itskovitz-Eldor et al. (2000)
Teratoma formation in vivo	+	+	Wobus et al. (1984), Thomson et al. (1998)

MEF mouse embryonic fibroblast, *EB* embryoid body

In vitro differentiation of mESCs requires the initial formation of 3D cell aggregates that have termed embryoid bodies (EBs), which may differentiate into into the three primordial embryonic germ layers. EBs are obtained by cultivating mESCs in hanging drops for 3 days and in suspension for further 2 days before being transferred onto gelatin-coated tissue culture dishes. The parameters that specifically influence the acquisition of cardiac phenotype are: (1) cell density in the EB, (2) culture medium, fetal bovine serum (FBS) or fetal calf serum (FCS), supplement with growth factors and other inducing agents (i.e. retinoic acid, DMSO, or co-colture with endoderm-like cells), (3) mESCs line, and (4) the time and duration of the differentiation process. Spontaneously, contracting cardiac myocytes become manifest between day 1 and 4 after plating between an outer epithelial layer and a basal layer of mesenchymal cells of the EB. The number of spontaneously beating foci and the rate of contraction rapidly augment through the differentiation process, but maturation (after about 1 month of culture) is accompanied by a reduction in average beating rate, as observed during normal cardiac development. Fully differentiated CMs normal fail to contract, but can be successfully maintained in culture for many weeks. Three stages can be distinguished during the differentiation process of murine EBs: early (pacemaker-like or primary myocardial-like cells), intermediate, and terminal (atrial-, ventricular-, nodal-, His-, and Purkinje-like cells) (Boheler et al. 2002; Becker et al. 2011). During early stages of differentiation, EBs-derived CMs (7+2 days) appear small and rounded, display sparse and irregularly organized myofibrils, or they can even lack them, and concentrate in round accumulations. At later stages (7+12 days), they organize into strands of elongated CMs endowed with mature sarcomeres and myofibrils (Hescheler et al. 1997; Boheler et al. 2002). Contracting myocytes are primarily mononuclear and rod-shaped cells, and exhibit cell-cell junctions resembling those that have been described in CMs of developing hearts. Densely packed and well organized myofibril bundles are evident at the terminal stages of differentiation, while sarcomeres show clear A bands, I bands, and Z disks. Likewise, intercalated disks, desmosomes, gap junctions, fascia adherens, and gap junctions have been detected, as well as the spreading of Lucifer yellow to adjoining cells after microinjection, which is indicative of functional gap junctions-mediated intercellular communication. Overall, the ultrastructural architecture, myofibrillar organization, and sarcomere length of mESCs-derived CMs are similar to those of foetal and neonatal rodent myocytes (Hescheler et al. 1997; Boheler et al. 2002).

CM differentiation of murine ESCs recapitulates the genetic program underlying cardiac development in mouse embryo in a developmentally regulated manner. The cardiac specific transcription factors Gata-4 and Nkx2.5 appear before transcripts encoding atrial natriuretic factor (ANF), myosin light chain (MLC)-2v, α-myosin heavy chain (α-MHC), β-myosin heavy chain (β-MHC), Na^+-Ca^{2+} exchanger and phospholamban. Similarly, sarcomeric proteins appear in the same time sequence as that described during cardiogenesis: titin (Z disk), α-actinin, myomesin, titin (M band), MHC, α-actin, cardiac troponin T, and M protein. Similar to foetal/neonatal CMs, early mESCs-derived CMs express slow skeletal muscle Troponin I and β-MHC rather α-MHC, which appear along with cardiac Troponin I in terminally differentiated CMs, displaying more rapid contractions (Boheler et al. 2002; Wobus and Boheler 2005). Concerning their electrophysiological properties, early (7+2 days) and intermediate (7+5−8 days) mESCs-derived CMs display spontaneous depolarizations, resembling the pacemaker-like action potentials of the sino-atrial node of fully differentiated heart. On the other hand, terminally differentiated (7+9−18 days) cells present three major types of action potential: spontaneous sinusoidal, triggered ventricular- and atrial-like action potentials. Moreover, terminal, but not early, mESCs-derived CMs respond to β-adrenergic stimulation. Finally, while T tubules are normally absent in these cells, the mechanism of excitation-contraction coupling resemble those found in foetal/neonatal CMs by utilizing ryanodine receptors (RyRs) to trigger Ca^{2+} release in response to membrane depolarization (Hescheler et al. 1997; Boheler et al. 2002; Moccia et al. 2014). In particular, the role of RyR2 progressively increases during cardiac differentiation, while Ca^{2+} mobilization during the early phases of development also involves inositol-1,4,5-trisphosphate receptors ($InsP_3Rs$) (Fu et al. 2006).

2.3 Human Embryonic Stem Cells Isolation, Characterization and CM Differentiation

The technique exploited to harvest and cultivate mESCs was adapted to generate hESC lines from discarded pre-implanted embryos produced by IVF. Thomson and collaborators accomplished the derivation of the first human ESC line through immunosurgery (Thomson et al. 1998). Thomson's group cultured human embryos up to the blastocyst stage and applied the immunosurgery technique (Solter and Knowles 1978) using a rabbit anti-rhesus spleen cell antiserum, that targets the external trophoectoderm cells, surrounding the ICM, and is responsible for the activation of the guinea pig complement. The final result is the release of the ICM. The ICM was plated and maintained for 16 days in suspension and then was grown for 3 weeks in gelatin until the new line was established: they obtained the first human ESC line, called H9.

The basic features of hESCs resemble those of their murine counterparts, such as high telomerase activity and teratoma formation in immunodeficient mice. The transcriptional pluripotency hub represented by Oct-3/4, Sox2 and Nanog is conserved in hESCs, but their target different signalling pathways as compared to mESCs, as reviewed in Schnerch et al. (2010). Additionally, hESCs show different isoforms of stage-specific antigens (SSEA-3 and SSEA-4) and proteoglycans (TRA-1–60, TRA-1–81, GCTM-2) that are absent in mESCs (Wobus and Boheler 2005). Similar to mESCs, hESCs form teratomas and teratocarcinomas in immunodeficient in vivo and actively proliferate for long periods of time in culture, but display longer average population doublings (30–35 *vs.* 12–15 h) (Amit et al. 2000). LIF does not maintain pluripotency in hESCs, which must be therefore grown on feeder layers of MEFs or human tissues-derived cells supplemented with bFGF2. Nevertheless, it has been found that hESCs remain pluripotent when cultured on

Matrigel or laminin in the presence of MEFs-conditioned media (Xu et al. 2001), albeit the soluble factor responsible for self-renewal maintenance has not been identified. As discussed below, hESCs are now cultivated on MEFs without animal serum (Lee et al. 2005) but in the presence of specific serum-free defined media (serum replacement) (Mummery et al. 2012) to avoid any xenogenic contamination that could hamper their subsequent application on human subjects.

HESCs-derived cardiomyocytes were described 3 years after hESC identification (Thomson et al. 1998; Kehat et al. 2001). HESCs (cell line H9.2) were dispersed into small clusters (3–20 cells) by using collagenase IV and then grown in suspension with serum for 7–10 days to form EBs, which were then coated onto gelatine-coated plates. Spontaneous contractions started 4 days after plating (i.e. 11–14 days after the beginning of the differentiation process), while the highest number of beating areas was detected 20 days after plating (i.e. after 27–30 days of differentiation) (Kehat et al. 2001). This protocol was thereafter utilized as the standard method to generate CMs from hESCs, but is flawed by several limitations. These include the low yield of CMs, which are as low as less than 1 % of the total cell population obtained by spontaneous hESC differentiation. Moreover, stimulating cell growth with FBS may not be the most suitable strategy for CBT because of the potential risks of contamination associated with serum. Innovative enrichment, purification and selection protocols have thus been established to drive cardiac differentiation toward relatively pure homogeneity. Defined culture media have been developed to improve the efficiency of CM derivation from hESCs. For instance, earlier studies demonstrated that co-culturing hESCs with END2, a mouse visceral endoderm-like cell-line, increased the percentage of contracting EBs to 35 % after 12 days in culture (Mummery et al. 2003). Subsequent studies showed that the cardiac inductive activity of END2 may be mimicked by culturing hESCs in insulin-free, serum-free medium supplemented with high concentrations of prostaglandin I2 (PGI2) (Xu et al. 2008a). Alternatively, specific growth factors involved in the initiation of the signalling pathways that control cardiogenesis can be used to favour CM generation from hESCs. For instance, activin 4 and BMP4, both members of the transforming growth factor β (TGFβ) family, may sequentially be added to the differentiation medium to favour efficient cardiac development. A monolayer of undifferentiated hESCs is plated on Matrigel before being exposed to activin A for 1 day and then to BMP5 for further 4 days, followed by removal of these growth factors and replacement with a serum-free medium for about 3 weeks. This protocol yields 30 % of CMs as compared to the conventional serum-induced differentiation (Laflamme et al. 2007). Alternatively, cardiac induction may be initiated by stimulating the Wnt/β-catenin signalling pathway, which controls cardiogenesis in zebrafish, *Xenopus* and mouse, with Wnt3 for 2 days in serum-containing medium, while activin A is added during the early phases of differentiation along with Wnt3 to accelerate CM formation. The final CM yield of this treatment is thus enhanced to ~50 % (Paige et al. 2010). Consistently, dickkopf homolog 1 (DKK1) adverses CM differentiation when is supplemented during the early stages of the protocol by inhibiting Wnt signalling (Tran et al. 2009). Nevertheless, CM generation is favoured when DKK1 is added at later stages (5–11 days) (Paige et al. 2010).

Finally, a staged protocol has been devised to recapitulate the microenvironment of developing mouse embryo by applying five growth factors at different time points during the differentiation process. Human ESCs are first stimulated with BMP4 for 1 day to form EBs and then challenged with BMP4, activin A, and basic fibroblast growth factor (bFGF) from days 1 to 4 to induce mesoderm. The development of cardiovascular lineages is promoted by the subsequent addition of vascular endothelial growth factor (VEGF), DKK1 and bFGF. The culture is maintained in hypoxic environment for the first 10–12 days and the final CM yield is ~40 % (Yang et al. 2008; Kattman et al. 2011).

The efficiency of CM differentiation might also be improved by targeting the signalling pathways involved in cardiogenesis by exploiting selective small molecule inhibitors or activators. For instance, low concentrations (10 μM) of SB203580, a specific inhibitor of p38 mitogen-activated protein kinase (MAPK), enhances CM formation from hESCs maintained in END2-conditioned medium by ~2 % (Graichen et al. 2008); the same effect is obtained when SB203580 is added to serum-free, insulin-free medium supplemented with high doses of PGI_2 (Xu et al. 2008a). Moreover, CM differentiation may be stimulated by treating hESCs with BMP2, another cardiogenic inducer, and SU5402, a FGF receptor blocker, albeit this effect is specific to well defined cell lines and culture conditions (Burridge et al. 2007; Yang et al. 2008). Finally, the demethylating agent 5-azacytidine, ascorbic acid, and cyclosporine A, but not retinoic acid and DMSO, were all reported to exert efficient cardiogenic effects (Passier et al. 2005; Xu et al. 2012).

The protocols illustrated above were conceived to direct cardiac differentiation from pluripotent hESCs; however, while the efficiency of CM differentiation is improved by such treatments, hESCs-derived cultures still contain cell types deriving from all three germ layers that might reduce the therapeutic outcome of CBT or even exert a pro-tumorigenic effect (Xu et al. 2012; Bernstein 2012). Several methods have been developed to enrich CMs from hESCs-derived cultures and enhance the purity of cell populations to be utilized for clinical applications. Manual dissection and dissociation of contracting areas from human ES cells-derived EBs could produce up to 70 % cells positively stained for cardiac markers (Mummery et al. 2003). A less labour intensive method is based on discontinuous Percoll gradient followed by centrifugation: most of CMs accumulate in the high percentage Percoll layer, thereby leading to three to sevenfold enrichment up to 70 % (Xu et al. 2002) and, eventually, 95 % CMs when differentiation is induced with activin A and BMP4 (Laflamme et al. 2007). Importantly, Percoll enriched CMs have successfully been employed in cardiac regeneration studies with no evidence of teratoma formation (Xu et al. 2012). A third strategy takes advantage from the high mitochondrial content of CMs as compared to non-cardiac cells. The fluorescent dye tetramethylrhodamine methyl ester perchlorate (TMRM) reversibly labels mitochondria and does not impair cell vitality. Cell sorting of differentiating human EBs carried out by exploiting TMRM fluorescence leads to a CM purity >90 %. When these cells were injected in immunodeficient mouse hearts, they showed a remarkable long-term (8 weeks) survival and did not form teratomas (Hattori et al. 2010). These enrichment protocols, although useful to obtain high purity CM populations from human EBs, present several limitations that could affect their therapeutic translation. None of them permit to obtain cardiac progenitor cells (CPCs) that could be useful for the reconstruction of injured myocardium in virtue of their ability to differentiate into multiple cells types, such as CMs, smooth muscle cells and endothelial cells. Moreover, mechanical dissection can only be performed towards the end of the differentiation process, when a sufficient number of contracting foci is detectable. Percoll gradient is also less efficient when is conducted at the beginning of human ES cell differentiation. Finally, TMRM purification is recommendable at the late stage of differentiation (from day 20 on), as at earlier stages many non-cardiac cells may be selected by this approach (Dubois et al. 2011; Elliott et al. 2011). Lineage-specific surface antigens have, therefore, been identified to selectively enrich CMs through fluorescence-activated or magnetic-assisted cell separation (FACS and MACS, respectively) procedures. Two cell surface cardiac markers have recently been associated to hESCs-derived CMs: vascular cell adhesion molecule 1 (VCAM1) and signal-regulatory protein alpha (SIRPA). FACS-based selection for $SIRPA^+$ cells leads to a population of 90–98 % cardiac Troponin T^+ ($cTnT^+$) cells (Dubois et al. 2011), while MACS-based sorting of $VCAM1^+$ cells results in a yield of 95 % $cTnT^+$ cells (Uosaki et al. 2011). Likewise, activated leukocyte cell adhesion molecule (ALCAM) is expressed in hESCs-derived CMs and is amenable to MACS. This approach leads to 58 % of the enriched population positive for both ALCAM and the cardiac marker α-MHC (Rust et al. 2009). These strategies present a number of hurdles as well. While FACS has a low throughput but may target multiple markers on individual cells, MACS displays a higher throughput but is limited to a single marker. Moreover, the percentage of hESCs-derived CMs expressing cTnT, SIRPA and VCAM1 amounts to only 37 % (Elliott et al. 2011), while it is still unknown whether ALCAM co-localizes with SIRPA and/or VCAM1 or identifies an additional CM population (Xu et al. 2012). It has, therefore, recently suggested to targeting non-cardiac markers to deplete non-cardiac cells and obtain pure CM-enriched cultures. Cell antigens exploited for this purpose include β-type platelet-derived growth factor receptor (PDGFRB; to exclude smooth muscle cells), platelet endothelial cell adhesion molecule-1 (PECAM-1; to exclude endothelial cells), and THY (to exclude fibroblasts) (Dubois et al. 2011).

Genetic selection provides the opportunity to purify hESCs-derived cardiomyocytes by placing a reporter gene or selectable marker under the control of a cardiac-specific promoter. This strategy, initially developed to obtain cultures consisting of >99.6 % CMs from mESCs (Klug et al. 1996), associated to the paucity of cell surface antigens has led to the development of several reporter hESC lines. Genetically-modified cells are thus induced to go through the differentiation process and, subsequently, differentiated CMs can be sorted by FACS, if the transgene encodes for a fluorescent protein (i.e. green fluorescent protein and mCherry), or selected by antibiotics (i.e. neomycin or geneticin), if the transgene determines antibiotic resistance (Wong and Bernstein 2010; Xu et al. 2012). Cardiac promoters that have been exploited to obtain purified CM culture include mouse or human cardiac α-MHC and human myosin light chain 2v (MLC2V). This strategy may yield up to 99 % of pure CMs that exhibit human adult cardiac genes and fire action potential similar to those recorded in fetal CMs (Kita-Matsuo et al. 2009; Wong and Bernstein 2010). Moreover, CMs generated through FACS or antibiotic resistance do not lead to teratoma formation (Xu et al. 2008a) and generate contractile forces similar to those of rat neonatal CMs (Kita-Matsuo et al. 2009). Nevertheless, their regenerative impact in the ischemic or failing heart remains to be elucidated; moreover, developing stable expression lines is an expensive and time consuming process that could be hampered by random integration of the selected transgene in the host genome, thereby causing genetic modifications of stem cells lines. This hurdle can be overcome by adopting the so-called dual fluorescence resonance energy transfer 'molecular beacon' (MB) technology for transient, real-time detection of gene expression during hESC differentiation. Molecular beacons are single-stranded oligonucleotide nanoscale probes with a fluorophore at the 5′ end and a quencher at the 3′ end. In the absence of their complementary target, MBs adopt a hairpin structure that causes the fluorophore to come in close contact with the quencher, thereby reducing the overall fluorescence. However, hybridization with a homologous sequence target opens the hairpin and separated the two opposite ends, thus leading to an increase in fluorescence. This strategy has been modified to detect the expression of specific cardiac mRNAs (i.e. Oct4 and MHC1) by using dual fluorescence resonance energy transfer (FRET) associated with FACS (King et al. 2011; Ban et al. 2013). Recently, MHC1-MBs$^+$ CMs have been shown to engraft up to 4 weeks in an AMI model without forming teratomas and significantly improving LVEF (by ~10 %) (Ban et al. 2013).

An extensive molecular characterization has demonstrated that hESCs-derived CMs express: (1) cardiac transcription factors, such as GATA-4, MEF-2, and Nkx2.5; (2) sarcomeric and myofibril proteins, such as α- and β-MHC, atrial and ventricular forms of MLC (MLC2A and MLC2V), tropomyosin, cTnT, cardiac Troponin I, α-actinin and desmin; and (3) metabolic regulators of cardiac development, such as ANF, myoglobin and creatine kinase-MB (Kehat et al. 2001; Xu et al. 2002). Nevertheless, hESCs-derived CMs present cytological and ultrastructural organization typical of foetal/neonatal, rather than adult, CMs: they are mononuclear cells, featured by a round, tri- or multi-angular shaped morphology, and with a high nuclear:cytoplasmic ratio. They, however, adopt a more elongated structure during maturation which is coupled to a progressive ultrastructural development from an irregular myofibrillar pattern to the generation of well-identifiable sarcomers at the end of the differentiation process. Accordingly, at this stage, some sarcomers show recognizable A-, I-, and Z-bands, albeit hESCs-derived CMs always lack T tubules (Snir et al. 2003; Passier et al. 2008). Likewise, hESCs-derived CMs display immature electrophysiological features. The beating rate in spontaneously contracting hEBs ranges between 30 and 130 bpm and undergoes positive and negative chronotropic regulation in response to adrenergic and cholinergic agonists, respectively. This stable pacemaker activity within hEBs is associated to functional gap junctions, comprising connexins 43 and 45, which enable the synchronized spreading of the electrophysiological signal through the cell aggregate. It must, however, be pointed out that: (1) connexin 45 expression is restricted at the early stages of in vivo development and (2) the beating activity may cease after the first

9 weeks of culture (Blazeski et al. 2012). Upon differentiation, hESCs-derived CMs exhibit a mixture of electrophysiological profiles that resemble those of the immature adult heart and are classified as nodal-, atrial-, and ventricular-like. This "functional signature" may vary depending on hESC lines, differentiation protocols and time of differentiation, as reviewed by Blazeski et al. (2012), and displays heterogeneous properties even among cell outgrowths deriving from the same EB. Consistent with the foetal/embryonal phenotype of the action potentials they trigger, hESCs-derived atrial- and ventricular-CMs are more depolarized (−40/−50 mV) than their adult counterparts (Passier et al. 2008; Blazeski et al. 2012). Moreover, the majority of hESCs-derived CMs shows a high degree of automaticity and spontaneously fires in absence of external stimuli; quiescent cells, however, trigger single action potentials upon stimulation and are, therefore, functionally active. The up-stroke is shaped by voltage-gated Na^+ channels (mainly encoded by $Na_V1.5$), but is significantly slower in ventricular hESCs-derived CMs than in adult cells (12 *vs.* 118 V/s). Ca^{2+} entry through L-type voltage-gated Ca^{2+} channels (encoded by $Ca_V1.2$) maintains the depolarization, while repolarization is accomplished by transient outward K^+ channels (I_{to}, encoded by Kv4.3), slow delayed rectifier K^+ channels (I_{Ks}, encoded by Kv7.1), and inward rectifier K^+ channels (I_{Kr}, encoded by Kv11.1 or human ether-a gò-gò related gene 1—HERG1). The spontaneous firing is initiated by voltage gated Na^+ channels with the contribution of the pacemaker current I_f (or funny current, mediated by hyperpolarization-activated cyclic nucleotide-gated channel 1 or HCN1). In general, most of these ion channels undergo a remarkable up-regulation over a long maturation period (i.e. 3–8 months) which moves the action potential shape toward a more adult phenotype, but I_f and HCN1 are down-regulated; this process is significantly slower as compared to the differentiation of mESCs-derived CMs which acquire an adult electrophysiological phenotype in approximately 3 weeks (Sartiani et al. 2007; Passier et al. 2008; Blazeski et al. 2012; Robertson et al. 2013). The mechanism of excitation-contraction coupling in hESCs-derived CMs is still highly debated. Overall, their intracellular Ca^{2+} signalling machinery is more immature as related to adult CMs, which show a positive force-frequency relationship due to the concerted interaction between voltage-operated Ca^{2+} inflow and RyRs activation at the dyadic junctions. Conversely, hESCs-derived CMs display a negative force-frequency, which reflects the lack of T-tubules and the immaturity of Ca^{2+} cycling across sarcoplasmic reticulum (SR) membranes. Furthermore, intracellular Ca^{2+} mobilization in these cells is dampened by a number of limiting factors, including: (1) low SR Ca^{2+} content, SERCA2A and RyR2 expression, which render them less sensitive to electrical stimulation and caffeine application; (2) absence of the SR regulatory proteins calsequestrin, junctin and triadin, which control SR Ca^{2+} content and RyR2 activation; (3) U-shaped propagation of electrically-induced intracellular Ca^{2+} waves. All of these features are typical of foetal/neonatal CMs and concur with the scarce contribution of RyRs2-induced Ca^{2+} release to the activation of the contractile machinery (Kong et al. 2010; Blazeski et al. 2012; Robertson et al. 2013). Consistently, abating RyRs2 signalling reduces the beating rate, but does not suppress contraction development. This immature phenotype is further supported by the observation that $InsP_3R2$ may be involved in intracellular Ca^{2+} mobilization and control of contraction rate (Jaconi et al. 2000; Robertson et al. 2013).

2.4 Advantages and Disadvantages of Embryonic Stem Cells as Source for Cardiac Regeneration

Among the different stem cell sources that can potentially be used to promote cardiac regeneration, hESCs stand out as the most promising candidates in virtue of their ability to differentiate into genuine CMs. Pioneering preclinical studies demonstrated that mESCs and mESCs-derived CMs may engraft and regenerate infracted myocardium by improving its contractile function (Christoforou and Gearhart 2007). Initially, hESCs-derived CMs were shown to restore the pacemaking activity in the heart of swines and guinea pigs with complete atrio-ventricular block or cryo-ablation, respectively (Passier et al. 2008). Later, it was demonstrated that hESCs-derived CMs survive, proliferate and mature for at least 12 weeks after injection into the healthy myocardium of immunodeficient mice or rats. However, grafted cells formed a syncytium that is separated from host tissue by a fibrotic patch secreted both by resident and transplanted CMs (Laflamme et al. 2005; Dai et al. 2007; Passier et al. 2008). Unfortunately, the engraftment success rate dropped from 90 to 18 % when hESCs-derived CMs were infused into a murine model of AMI. The same study disclosed that a pro-survival cocktail, containing components of the extracellular matrix and anti-death factors, improved graft survival and, consequently, ventricular contractility up to 4 weeks post-transplantation (Laflamme et al. 2005). Additional studies confirmed that hESCs-derived CMs represent a feasible tool to induce cardiac repair in rodent models of AMI (van Laake et al. 2007a, b, 2008; Leor et al. 2007; Caspi et al. 2007a). Nevertheless, no clinical trials have been launched yet as a consequence of the major challenges that remain to be overcome before the clinical translation of hESCs-derived CMs. First, the use of human-derived cells has raised serious ethical and legal concerns and it does not easily gain the appreciation of the public opinion (Joggerst and Hatzopoulos 2009; Wong and Bernstein 2010; Jakob and Landmesser 2013; Pfister et al. 2014). Second, functional experiments unveiled that their effect on cardiac performance is only transient. For instance, even when the dose of hESCs-derived CM injected into injured heart is tripled, no functional improvement is observed after 12 weeks (van Laake et al. 2007a, b, 2008). This therapeutically relevant hurdle might be solved by using hESCs-derived cardiac progenitor cells (CPCs), which possess a superior ability to integrate within the injured heart and restore cardiac vascularisation (Wong and Bernstein 2010; Bernstein 2012). However, the transient improvement of ventricular function induced by hESCs-derived CMs may also reflect the mismatch between the beating rates of human *vs.* rodent CMs (60–100 *vs.* 300–600 bpm). This would reflect in the functional uncoupling that leads to the failure of long-term treatment (Passier et al. 2008). Therefore, rodent hearts might not provide the most suitable model to study the regenerative potential of hESCs, and larger animals, such as pigs and primates, with beating rates comparable to those of humans could be used to address this issue. Third, hESCs may give raise to teratocarcinomas at the implantation site. Albeit these teratomas are regarded as benign in vivo, some studies reported the expression of malignant markers in some cells. The injection of hESCs-derived

CMs is predicted to alleviate this side effect, but these findings raise ethical concerns about the safety of their clinical use. Fourth, allogenic hESCs are obtained from embryos and obviously do not retain the same genome as the patients. It turns out that, although they are immunologically immature, hESCs-based therapy requires an immunosuppressive regimen (i.e. tacrolimus or cyclosporine) to prevent immune rejection after transplantation. Induced pluripotent stem (iPS) cells may be generated by reprogramming adult autologous differentiated cells (Takahashi and Yamanaka 2006; Takahashi et al. 2007) and then directed to differentiate towards CMs. Such strategy circumvents the limitation imposed by ethical concerns and immunological reaction, but is extremely expensive, time consuming and not exempt by risk (teratoma formation in vivo, virus as vectors, genomic rearrangements and epigenetic mutations) (Bernstein 2012; Akhmedov and Marín-García 2013). However, protocols for cardiac differentiation of hESCs have been significantly improved to attenuate the risk of teratoma formation, favour stem cell engraftment, and increase functional improvement (Mummery et al. 2012; Bernstein 2012). For instance, alternative culture products and feeder-free culture systems have been devised to obtain and maintain hESCs under animal- and xenobiotic-free conditions (Bernstein 2012). Moreover, earlier hESC lines have been discarded for clinical use as long-term cultures lead to chromosomic and genomic instabilities; therefore, new hESC lines are required that should be maintained under conditions that do not compromise genomic integrity improvement (Mummery et al. 2012; Bernstein 2012). This goal could, for instance, be achieved by exploiting culture media that induce adult-like energy metabolism (Maher and Xu 2013).

2.5 Future Trends in the Therapeutic Application of hESCs for Cardiac Regeneration

In the next future, unveiling the molecular determinants of cardiogenesis will favour the development of more robust and reliable methodologies to promote hESCs differentiation into cardiac myocytes. Recent work clearly showed that miR-1 and miR-133 are enriched in mESCs and hESCs-derived CMs and are involved in the acquisition of a cardiac phenotype. Intriguingly, miR-1 over-expression increases Nkx2-5 expression in hESCs and results in more than a threefold higher number of beating hEBs as compared with wild-type cells (Ivey et al. 2008). Epigenetic manipulation could provide an additional means to control stem cell fate. The histone deacetylase inhibitor trichostatin A (TSA) may be administrated to mESCs to favor cardiogenesis by enhancing the expression of Nkx2-5, ANF, and β-MHC. Moreover, mESCs may be committed toward the cardiac lineage by the ectopic expression of Baf60c, a cardiac-enriched subunit of the Swi/Snf-like BAF chromatin remodeling complex, and the cardiac transcription factors GATA4 and Tbx5 (Takeuchi and Bruneau 2009). These observations remain to be confirmed in hESCs, but pave the way to design alternative protocols to promote cardiac differentiation for regenerative purposes. In this context, it will be important to understand how specifically directing the differentiation process towards the nodal or working (atrial and ventricular) phenotype. Currently available protocols do not distinguish between these two different cardiac subtypes which might induce undesired off-target effects in recipient hearts. While working-type cardiomyocytes are more suitable for cardiac regeneration, enriched nodal-type cardiomyocytes are more amenable for restoring the pacemaker function in arrhythmic patients. Contamination with nodal-type cells might thus enhance the risk for arrhythmia in ischemic/failing patients and represents a potential risk for hESCs-based therapy of CD (Maher and Xu 2013). Profiling the panel of cell surface antigens that identify cardiac progenitor pools and improving non-genetic selection strategies will permit to enrich the hESCs-derived population of CMs ready for clinical use. Recent advances in tissue engineering are predicted to enhance long-term graft and survival of hESCs-derived CMs. A number of 3D human cardiac patches have been constructed that contain embryonic fibroblasts, human endothelial cells, and hESCs-derived CMs (Caspi et al. 2007b; Lesman et al. 2010). This multicellular preparation is fully vascularized and capable of producing spontaneous and synchronized contractions that are propagated by gap junction connecting adjoining CMs (Caspi et al. 2007b). Relevant to reparative purposes, such tissue-engineered human vascularized cardiac muscle forms stable biografts in the heart of recipient rats (Lesman et al. 2010). It will be important to develop a reliable and reproducible strategy to prevent immune rejection of injected cells. Preliminary studies conducted on patients suffering from spinal cord injury and receiving hESCs-derived oligodendrocytes did not show any deleterious consequence after 6 months of moderate immunosuppression. The lower drug doses employed, as compared to whole organ transplantation, are justified by the lack of antigen presenting cells within the cellular graft, which reduces its immunogenicity (Menasché 2012). Finally, hESCs might contribute to cardiac regeneration by means other than cell replacement, i.e. by paracrine release of cardioprotective mediators. The injection of hESCs-derived CMs in a mouse model of AMI has been shown to increase angiogenesis, reduce apoptosis, infarct scar and myocardial fibrosis, and improve LVEF even in the absence of long-term engraftment. This observation suggests that hESCs-derived CMs secrete soluble factors with the potential to stimulate the endogenous mechanisms of cardiac healing. The identification and isolation of such biofactors could lead to the synthesis of "off-the-shelf" drugs to restore myocardial structure and function with no need of CBT and circumventing all its associated hurdles and concerns.

References

Akhmedov AT, Marín-García J (2013) Myocardial regeneration of the failing heart. Heart Fail Rev 18:815–833. doi:10.1007/s10741-012-9348-5

Amit M, Carpenter MK, Inokuma MS et al (2000) Clonally derived human embryonic stem cell lines maintain pluripotency and proliferative potential for prolonged periods of culture. Dev Biol 227:271–278. doi:10.1006/dbio.2000.9912

Anderson D, Self T, Mellor IR et al (2007) Transgenic enrichment of cardiomyocytes from human embryonic stem cells. Mol Ther 15:2027–2036. doi:10.1038/sj.mt.6300303

Armstrong L, Lako M, Lincoln J et al (2000) mTert expression correlates with telomerase activity during the differentiation of murine embryonic stem cells. Mech Dev 97:109–116
Avilion AA, Nicolis SK, Pevny LH et al (2003) Multipotent cell lineages in early mouse development depend on SOX2 function. Genes Dev 17:126–140. doi:10.1101/gad.224503
Bader A, Al-Dubai H, Weitzer G (2000) Leukemia inhibitory factor modulates cardiogenesis in embryoid bodies in opposite fashions. Circ Res 86:787–794
Bader A, Gruss A, Höllrigl A et al (2001) Paracrine promotion of cardiomyogenesis in embryoid bodies by LIF modulated endoderm. Differentiation 68:31–43. doi:10.1046/j.1432-0436.2001.068001031.x
Ban K, Wile B, Kim S et al (2013) Purification of cardiomyocytes from differentiating pluripotent stem cells using molecular beacons that target cardiomyocyte-specific mRNA. Circulation 128:1897–1909. doi:10.1161/CIRCULATIONAHA.113.004228
Becker JR, Deo RC, Werdich AA et al (2011) Human cardiomyopathy mutations induce myocyte hyperplasia and activate hypertrophic pathways during cardiogenesis in zebrafish. Dis Model Mech 4:400–410. doi:10.1242/dmm.006148
Behfar A, Zingman LV, Hodgson DM et al (2002) Stem cell differentiation requires a paracrine pathway in the heart. FASEB J 16:1558–1566. doi:10.1096/fj.02-0072com
Bernstein HS (2012) Cardiac repair and restoration using human embryonic stem cells. Regen Med 7:697–712. doi:10.2217/rme.12.46
Blazeski A, Zhu R, Hunter DW et al (2012) Electrophysiological and contractile function of cardiomyocytes derived from human embryonic stem cells. Prog Biophys Mol Biol 110:178–195. doi:10.1016/j.pbiomolbio.2012.07.012
Boheler KR, Czyz J, Tweedie D et al (2002) Differentiation of pluripotent embryonic stem cells into cardiomyocytes. Circ Res 91:189–201
Buggisch M, Ateghang B, Ruhe C et al (2007) Stimulation of ES-cell-derived cardiomyogenesis and neonatal cardiac cell proliferation by reactive oxygen species and NADPH oxidase. J Cell Sci 120:885–894. doi:10.1242/jcs.03386
Burridge PW, Anderson D, Priddle H et al (2007) Improved human embryonic stem cell embryoid body homogeneity and cardiomyocyte differentiation from a novel V-96 plate aggregation system highlights interline variability. Stem Cells 25:929–938. doi:10.1634/stemcells.2006-0598
Carpenter MK, Rosler ES, Fisk GJ et al (2004) Properties of four human embryonic stem cell lines maintained in a feeder-free culture system. Dev Dyn 229:243–258. doi:10.1002/dvdy.10431
Caspi O, Huber I, Kehat I et al (2007a) Transplantation of human embryonic stem cell-derived cardiomyocytes improves myocardial performance in infarcted rat hearts. J Am Coll Cardiol 50:1884–1893. doi:10.1016/j.jacc.2007.07.054
Caspi O, Lesman A, Basevitch Y et al (2007b) Tissue engineering of vascularized cardiac muscle from human embryonic stem cells. Circ Res 100:263–272. doi:10.1161/01.RES.0000257776.05673.ff
Chambers I, Colby D, Robertson M et al (2003) Functional expression cloning of Nanog, a pluripotency sustaining factor in embryonic stem cells. Cell 113:643–655
Christoforou N, Gearhart JD (2007) Stem cells and their potential in cell-based cardiac therapies. Prog Cardiovasc Dis 49:396–413. doi:10.1016/j.pcad.2007.02.006
Dai W, Field LJ, Rubart M et al (2007) Survival and maturation of human embryonic stem cell-derived cardiomyocytes in rat hearts. J Mol Cell Cardiol 43:504–516. doi:10.1016/j.yjmcc.2007.07.001
Doetschman TC, Eistetter H, Katz M et al (1985) The in vitro development of blastocyst-derived embryonic stem cell lines: formation of visceral yolk sac, blood islands and myocardium. J Embryol Exp Morphol 87:27–45
Dubois NC, Craft AM, Sharma P et al (2011) SIRPA is a specific cell-surface marker for isolating cardiomyocytes derived from human pluripotent stem cells. Nat Biotechnol 29:1011–1018. doi:10.1038/nbt.2005
Elliott DA, Braam SR, Koutsis K et al (2011) NKX2-5(eGFP/w) hESCs for isolation of human cardiac progenitors and cardiomyocytes. Nat Methods 8:1037–1040. doi:10.1038/nmeth.1740
Evans MJ, Kaufman MH (1981) Establishment in culture of pluripotential cells from mouse embryos. Nature 292:154–156
Fu J, Yu H, Wang R et al (2006) Developmental regulation of intracellular calcium transients during cardiomyocyte differentiation of mouse embryonic stem cells. Acta Pharmacol Sin 27:901–910. doi:10.1111/j.1745-7254.2006.00380.x

Gaztelumendi N, Nogués C (2014) Chromosome instability in mouse embryonic stem cells. Sci Rep 4:5324. doi:10.1038/srep05324
Ginis I, Luo Y, Miura T et al (2004) Differences between human and mouse embryonic stem cells. Dev Biol 269:360–380. doi:10.1016/j.ydbio.2003.12.034
Graf U, Casanova EA, Cinelli P (2011) The role of the leukemia inhibitory factor (LIF)—pathway in derivation and maintenance of murine pluripotent stem cells. Genes (Basel) 2:280–297. doi:10.3390/genes2010280
Graichen R, Xu X, Braam SR et al (2008) Enhanced cardiomyogenesis of human embryonic stem cells by a small molecular inhibitor of p38 MAPK. Differentiation 76:357–370. doi:10.1111/j.1432-0436.2007.00236.x
Guan K, Fürst DO, Wobus AM (1999) Modulation of sarcomere organization during embryonic stem cell-derived cardiomyocyte differentiation. Eur J Cell Biol 78:813–823. doi:10.1016/S0171-9335(99)80032-6
Gwak S-J, Bhang SH, Kim I-K et al (2008) The effect of cyclic strain on embryonic stem cell-derived cardiomyocytes. Biomaterials 29:844–856. doi:10.1016/j.biomaterials.2007.10.050
Hattori F, Chen H, Yamashita H et al (2010) Nongenetic method for purifying stem cell-derived cardiomyocytes. Nat Methods 7:61–66. doi:10.1038/nmeth.1403
Henderson JK, Draper JS, Baillie HS et al (2002) Preimplantation human embryos and embryonic stem cells show comparable expression of stage-specific embryonic antigens. Stem Cells 20:329–337. doi:10.1634/stemcells.20-4-329
Hescheler J, Fleischmann BK, Lentini S et al (1997) Embryonic stem cells: a model to study structural and functional properties in cardiomyogenesis. Cardiovasc Res 36:149–162
Huh NE, Pasumarthi KB, Soonpaa MH et al (2001) Functional abrogation of p53 is required for T-Ag induced proliferation in cardiomyocytes. J Mol Cell Cardiol 33:1405–1419. doi:10.1006/jmcc.2001.1403
Itskovitz-Eldor J, Schuldiner M, Karsenti D et al (2000) Differentiation of human embryonic stem cells into embryoid bodies compromising the three embryonic germ layers. Mol Med 6:88–95
Ivey KN, Muth A, Arnold J et al (2008) MicroRNA regulation of cell lineages in mouse and human embryonic stem cells. Cell Stem Cell 2:219–229. doi:10.1016/j.stem.2008.01.016
Jaconi M, Bony C, Richards SM et al (2000) Inositol 1,4,5-trisphosphate directs Ca(2+) flow between mitochondria and the Endoplasmic/Sarcoplasmic reticulum: a role in regulating cardiac autonomic Ca(2+) spiking. Mol Biol Cell 11:1845–1858
Jakob P, Landmesser U (2013) Current status of cell-based therapy for heart failure. Curr Heart Fail Rep 10:165–176. doi:10.1007/s11897-013-0134-z
Joggerst SJ, Hatzopoulos AK (2009) Stem cell therapy for cardiac repair: benefits and barriers. Expert Rev Mol Med 11, e20. doi:10.1017/S1462399409001124
Kania G, Corbeil D, Fuchs J et al (2005) Somatic stem cell marker prominin-1/CD133 is expressed in embryonic stem cell-derived progenitors. Stem Cells 23:791–804. doi:10.1634/stemcells.2004-0232
Kattman SJ, Huber TL, Keller GM (2006) Multipotent flk-1+ cardiovascular progenitor cells give rise to the cardiomyocyte, endothelial, and vascular smooth muscle lineages. Dev Cell 11:723–732. doi:10.1016/j.devcel.2006.10.002
Kattman SJ, Witty AD, Gagliardi M et al (2011) Stage-specific optimization of activin/nodal and BMP signaling promotes cardiac differentiation of mouse and human pluripotent stem cell lines. Cell Stem Cell 8:228–240. doi:10.1016/j.stem.2010.12.008
Kehat I, Kenyagin-Karsenti D, Snir M et al (2001) Human embryonic stem cells can differentiate into myocytes with structural and functional properties of cardiomyocytes. J Clin Invest 108:407–414. doi:10.1172/JCI12131
Kehat I, Khimovich L, Caspi O et al (2004) Electromechanical integration of cardiomyocytes derived from human embryonic stem cells. Nat Biotechnol 22:1282–1289. doi:10.1038/nbt1014
Kim HS, Cho JW, Hidaka K, Morisaki T (2007) Activation of MEK-ERK by heregulin-beta1 promotes the development of cardiomyocytes derived from ES cells. Biochem Biophys Res Commun 361:732–738. doi:10.1016/j.bbrc.2007.07.045

King FW, Liszewski W, Ritner C, Bernstein HS (2011) High-throughput tracking of pluripotent human embryonic stem cells with dual fluorescence resonance energy transfer molecular beacons. Stem Cells Dev 20:475–484. doi:10.1089/scd.2010.0219

Kita-Matsuo H, Barcova M, Prigozhina N et al (2009) Lentiviral vectors and protocols for creation of stable hESC lines for fluorescent tracking and drug resistance selection of cardiomyocytes. PLoS One 4, e5046. doi:10.1371/journal.pone.0005046

Klug MG, Soonpaa MH, Koh GY, Field LJ (1996) Genetically selected cardiomyocytes from differentiating embronic stem cells form stable intracardiac grafts. J Clin Invest 98:216–224. doi:10.1172/JCI118769

Kong C-W, Akar FG, Li RA (2010) Translational potential of human embryonic and induced pluripotent stem cells for myocardial repair: insights from experimental models. Thromb Haemost 104:30–38. doi:10.1160/TH10-03-0189

Laflamme MA, Gold J, Xu C et al (2005) Formation of human myocardium in the rat heart from human embryonic stem cells. Am J Pathol 167:663–671. doi:10.1016/S0002-9440(10)62041-X

Laflamme MA, Chen KY, Naumova AV et al (2007) Cardiomyocytes derived from human embryonic stem cells in pro-survival factors enhance function of infarcted rat hearts. Nat Biotechnol 25:1015–1024. doi:10.1038/nbt1327

Lee JB, Lee JE, Park JH et al (2005) Establishment and maintenance of human embryonic stem cell lines on human feeder cells derived from uterine endometrium under serum-free condition. Biol Reprod 72:42–49. doi:10.1095/biolreprod.104.033480

Leor J, Gerecht S, Cohen S et al (2007) Human embryonic stem cell transplantation to repair the infarcted myocardium. Heart 93:1278–1284. doi:10.1136/hrt.2006.093161

Lesman A, Habib M, Caspi O et al (2010) Transplantation of a tissue-engineered human vascularized cardiac muscle. Tissue Eng Part A 16:115–125. doi:10.1089/ten.TEA.2009.0130

Maher KO, Xu C (2013) Marching towards regenerative cardiac therapy with human pluripotent stem cells. Discov Med 15:349–356

Martin GR (1981) Isolation of a pluripotent cell line from early mouse embryos cultured in medium conditioned by teratocarcinoma stem cells. Proc Natl Acad Sci U S A 78:7634–7638

Menasché P (2012) Embryonic stem cells for severe heart failure: why and how? J Cardiovasc Transl Res 5:555–565. doi:10.1007/s12265-012-9356-9

Mitsui K, Tokuzawa Y, Itoh H et al (2003) The homeoprotein Nanog is required for maintenance of pluripotency in mouse epiblast and ES cells. Cell 113:631–642

Moccia F, Dragoni S, Cinelli M et al (2013) How to utilize Ca2+ signals to rejuvenate the repairative phenotype of senescent endothelial progenitor cells in elderly patients affected by cardiovascular diseases: a useful therapeutic support of surgical approach? BMC Surg 13(suppl 2):S46. doi:10.1186/1471-2482-13-S2-S46

Moccia F, Ruffinatti FA, Zuccolo E (2014) Intracellular Ca2+ signals to reconstruct a broken heart: still a theoretical approach? Curr Drug Targets 16(8):793–815

Müller M, Fleischmann BK, Selbert S et al (2000) Selection of ventricular-like cardiomyocytes from ES cells in vitro. FASEB J 14:2540–2548. doi:10.1096/fj.00-0002com

Mummery C, Ward-van Oostwaard D, Doevendans P et al (2003) Differentiation of human embryonic stem cells to cardiomyocytes: role of coculture with visceral endoderm-like cells. Circulation 107:2733–2740. doi:10.1161/01.CIR.0000068356.38592.68

Mummery CL, Ward D, Passier R (2007) Differentiation of human embryonic stem cells to cardiomyocytes by coculture with endoderm in serum-free medium. Curr Protoc Stem Cell Biol Chapter 1:Unit 1F.2. doi: 10.1002/9780470151808.sc01f02s2

Mummery CL, Zhang J, Ng ES et al (2012) Differentiation of human embryonic stem cells and induced pluripotent stem cells to cardiomyocytes: a methods overview. Circ Res 111:344–358. doi:10.1161/CIRCRESAHA.110.227512

Neri T, Monti M, Rebuzzini P et al (2007) Mouse fibroblasts are reprogrammed to Oct-4 and Rex-1 gene expression and alkaline phosphatase activity by embryonic stem cell extracts. Cloning Stem Cells 9:394–406. doi:10.1089/clo.2006.0011

Niwa H, Burdon T, Chambers I, Smith A (1998) Self-renewal of pluripotent embryonic stem cells is mediated via activation of STAT3. Genes Dev 12:2048–2060

Paige SL, Osugi T, Afanasiev OK et al (2010) Endogenous Wnt/beta-catenin signaling is required for cardiac differentiation in human embryonic stem cells. PLoS One 5, e11134. doi:10.1371/journal.pone.0011134
Parsons XH (2012) Mending the broken heart—towards clinical application of human embryonic stem cell therapy derivatives. J Clin Exp Cardiol 3, e116
Passier R, Denning C, Mummery C (2006) Cardiomyocytes from human embryonic stem cells. Handb Exp Pharmacol 174:101–122
Passier R, van Laake LW, Mummery CL (2008) Stem-cell-based therapy and lessons from the heart. Nature 453:322–329. doi:10.1038/nature07040
Pasumarthi KB, Tsai SC, Field LJ (2001) Coexpression of mutant p53 and p193 renders embryonic stem cell-derived cardiomyocytes responsive to the growth-promoting activities of adenoviral E1A. Circ Res 88:1004–1011
Pera MF, Reubinoff B, Trounson A (2000) Human embryonic stem cells. J Cell Sci 113(pt 1):5–10
Pesce M, Anastassiadis K, Schöler HR (1999) Oct-4: lessons of totipotency from embryonic stem cells. Cells Tissues Organs 165:144–152
Pfister O, Della Verde G, Liao R, Kuster GM (2014) Regenerative therapy for cardiovascular disease. Transl Res 163:307–320. doi:10.1016/j.trsl.2013.12.005
Rebuzzini P, Martinelli P, Blasco M et al (2007) Inhibition of gene amplification in telomerase deficient immortalized mouse embryonic fibroblasts. Carcinogenesis 28:553–559. doi:10.1093/carcin/bgl158
Reubinoff BE, Pera MF, Fong CY et al (2000) Embryonic stem cell lines from human blastocysts: somatic differentiation in vitro. Nat Biotechnol 18:399–404. doi:10.1038/74447
Richards M, Tan S-P, Tan J-H et al (2004) The transcriptome profile of human embryonic stem cells as defined by SAGE. Stem Cells 22:51–64. doi:10.1634/stemcells.22-1-51
Robertson C, Tran DD, George SC (2013) Concise review: maturation phases of human pluripotent stem cell-derived cardiomyocytes. Stem Cells 31:829–837. doi:10.1002/stem.1331
Rudy-Reil D, Lough J (2004) Avian precardiac endoderm/mesoderm induces cardiac myocyte differentiation in murine embryonic stem cells. Circ Res 94:e107–e116. doi:10.1161/01.RES.0000134852.12783.6e
Rust W, Balakrishnan T, Zweigerdt R (2009) Cardiomyocyte enrichment from human embryonic stem cell cultures by selection of ALCAM surface expression. Regen Med 4:225–237. doi:10.2217/17460751.4.2.225
Sachinidis A, Schwengberg S, Hippler-Altenburg R et al (2006) Identification of small signalling molecules promoting cardiac-specific differentiation of mouse embryonic stem cells. Cell Physiol Biochem 18:303–314. doi:10.1159/000097608
Sartiani L, Bettiol E, Stillitano F et al (2007) Developmental changes in cardiomyocytes differentiated from human embryonic stem cells: a molecular and electrophysiological approach. Stem Cells 25:1136–1144. doi:10.1634/stemcells.2006-0466
Sauer H, Rahimi G, Hescheler J, Wartenberg M (1999) Effects of electrical fields on cardiomyocyte differentiation of embryonic stem cells. J Cell Biochem 75:710–723
Sauer H, Rahimi G, Hescheler J, Wartenberg M (2000) Role of reactive oxygen species and phosphatidylinositol 3-kinase in cardiomyocyte differentiation of embryonic stem cells. FEBS Lett 476:218–223
Schnerch A, Cerdan C, Bhatia M (2010) Distinguishing between mouse and human pluripotent stem cell regulation: the best laid plans of mice and men. Stem Cells 28:419–430. doi:10.1002/stem.298
Schroeder M, Niebruegge S, Werner A et al (2005) Differentiation and lineage selection of mouse embryonic stem cells in a stirred bench scale bioreactor with automated process control. Biotechnol Bioeng 92:920–933. doi:10.1002/bit.20668
Sharifpanah F, Wartenberg M, Hannig M et al (2008) Peroxisome proliferator-activated receptor alpha agonists enhance cardiomyogenesis of mouse ES cells by utilization of a reactive oxygen species-dependent mechanism. Stem Cells 26:64–71. doi:10.1634/stemcells.2007-0532

Shimko VF, Claycomb WC (2008) Effect of mechanical loading on three-dimensional cultures of embryonic stem cell-derived cardiomyocytes. Tissue Eng Part A 14:49–58. doi:10.1089/ten.a.2007.0092

Singh AM, Li F-Q, Hamazaki T et al (2007) Chibby, an antagonist of the Wnt/beta-catenin pathway, facilitates cardiomyocyte differentiation of murine embryonic stem cells. Circulation 115:617–626. doi:10.1161/CIRCULATIONAHA.106.642298

Smith AG, Heath JK, Donaldson DD et al (1988) Inhibition of pluripotential embryonic stem cell differentiation by purified polypeptides. Nature 336:688–690. doi:10.1038/336688a0

Snir M, Kehat I, Gepstein A et al (2003) Assessment of the ultrastructural and proliferative properties of human embryonic stem cell-derived cardiomyocytes. Am J Physiol Heart Circ Physiol 285:H2355–H2363. doi:10.1152/ajpheart.00020.2003

Solter D, Knowles BB (1978) Monoclonal antibody defining a stage-specific mouse embryonic antigen (SSEA-1). Proc Natl Acad Sci U S A 75:5565–5569

Sudou A, Muramatsu H, Kaname T et al (1997) Le(X) structure enhances myocardial differentiation from embryonic stem cells. Cell Struct Funct 22:247–251

Takahashi K, Yamanaka S (2006) Induction of pluripotent stem cells from mouse embryonic and adult fibroblast cultures by defined factors. Cell 126:663–676. doi:10.1016/j.cell.2006.07.024

Takahashi T, Lord B, Schulze PC et al (2003) Ascorbic acid enhances differentiation of embryonic stem cells into cardiac myocytes. Circulation 107:1912–1916. doi:10.1161/01.CIR.0000064899.53876.A3

Takahashi K, Tanabe K, Ohnuki M et al (2007) Induction of pluripotent stem cells from adult human fibroblasts by defined factors. Cell 131:861–872. doi:10.1016/j.cell.2007.11.019

Takeuchi JK, Bruneau BG (2009) Directed transdifferentiation of mouse mesoderm to heart tissue by defined factors. Nature 459:708–711. doi:10.1038/nature08039

Thomson JA, Itskovitz-Eldor J, Shapiro SS et al (1998) Embryonic stem cell lines derived from human blastocysts. Science 282:1145–1147

Tran TH, Wang X, Browne C et al (2009) Wnt3a-induced mesoderm formation and cardiomyogenesis in human embryonic stem cells. Stem Cells 27:1869–1878. doi:10.1002/stem.95

Uosaki H, Fukushima H, Takeuchi A et al (2011) Efficient and scalable purification of cardiomyocytes from human embryonic and induced pluripotent stem cells by VCAM1 surface expression. PLoS One 6, e23657. doi:10.1371/journal.pone.0023657

Van Laake LW, Passier R, Monshouwer-Kloots J et al (2007a) Monitoring of cell therapy and assessment of cardiac function using magnetic resonance imaging in a mouse model of myocardial infarction. Nat Protoc 2:2551–2567. doi:10.1038/nprot.2007.371

Van Laake LW, Passier R, Monshouwer-Kloots J et al (2007b) Human embryonic stem cell-derived cardiomyocytes survive and mature in the mouse heart and transiently improve function after myocardial infarction. Stem Cell Res 1:9–24. doi:10.1016/j.scr.2007.06.001

Van Laake LW, Passier R, Doevendans PA, Mummery CL (2008) Human embryonic stem cell-derived cardiomyocytes and cardiac repair in rodents. Circ Res 102:1008–1010. doi:10.1161/CIRCRESAHA.108.175505

Wo Y, Zhu D, Hu Y et al (2008) Reactive oxygen species involved in prenylflavonoids, icariin and icaritin, initiating cardiac differentiation of mouse embryonic stem cells. J Cell Biochem 103:1536–1550. doi:10.1002/jcb.21541

Wobus AM, Boheler KR (2005) Embryonic stem cells: prospects for developmental biology and cell therapy. Physiol Rev 85:635–678. doi:10.1152/physrev.00054.2003

Wobus AM, Holzhausen H, Jäkel P, Schöneich J (1984) Characterization of a pluripotent stem cell line derived from a mouse embryo. Exp Cell Res 152:212–219

Wobus AM, Kaomei G, Shan J et al (1997) Retinoic acid accelerates embryonic stem cell-derived cardiac differentiation and enhances development of ventricular cardiomyocytes. J Mol Cell Cardiol 29:1525–1539

Wong SSY, Bernstein HS (2010) Cardiac regeneration using human embryonic stem cells: producing cells for future therapy. Regen Med 5:763–775. doi:10.2217/rme.10.52

Xu C, Inokuma MS, Denham J et al (2001) Feeder-free growth of undifferentiated human embryonic stem cells. Nat Biotechnol 19:971–974. doi:10.1038/nbt1001-971

Xu C, Police S, Rao N, Carpenter MK (2002) Characterization and enrichment of cardiomyocytes derived from human embryonic stem cells. Circ Res 91:501–508

Xu C, Police S, Hassanipour M, Gold JD (2006) Cardiac bodies: a novel culture method for enrichment of cardiomyocytes derived from human embryonic stem cells. Stem Cells Dev 15:631–639. doi:10.1089/scd.2006.15.631

Xu XQ, Graichen R, Soo SY et al (2008a) Chemically defined medium supporting cardiomyocyte differentiation of human embryonic stem cells. Differentiation 76:958–970. doi:10.1111/j.1432-0436.2008.00284.x

Xu XQ, Zweigerdt R, Soo SY et al (2008b) Highly enriched cardiomyocytes from human embryonic stem cells. Cytotherapy 10:376–389. doi:10.1080/14653240802105307

Xu H, Yi BA, Wu H et al (2012) Highly efficient derivation of ventricular cardiomyocytes from induced pluripotent stem cells with a distinct epigenetic signature. Cell Res 22:142–154. doi:10.1038/cr.2011.171

Yang L, Soonpaa MH, Adler ED et al (2008) Human cardiovascular progenitor cells develop from a KDR+ embryonic-stem-cell-derived population. Nature 453:524–528. doi:10.1038/nature06894

Ying QL, Nichols J, Chambers I, Smith A (2003) BMP induction of Id proteins suppresses differentiation and sustains embryonic stem cell self-renewal in collaboration with STAT3. Cell 115:281–292

Zandstra PW, Bauwens C, Yin T et al (2003) Scalable production of embryonic stem cell-derived cardiomyocytes. Tissue Eng 9:767–778. doi:10.1089/107632703768247449

Chapter 3
Induced Pluripotent Stem Cells for Cardiac Regeneration

Rosalinda Madonna, Peter Ferdinandy, and Rainer Schulz

3.1 Human Induced Pluripotent Stem Cells

Although therapies using human embryonic stem cell (hESCs) have shown greatest potential that lies in their pluripotency, many ethical, biological and safety issues remain before they can be used in the clinical setting. One of the main concerns about transplanting hESCs is the possibility of immunological rejection. The development of "self" stem cells that are able to form all three germ layers—the endoderm (which becomes the liver and pancreas), the mesoderm (heart muscle) and the ectoderm (neural tissue), could be a strategy to overcome the limitations of hESCs. Strategies for the development of self pluripotent stem cells include somatic cell nuclear transfer and generation of induced pluripotent stem (iPS) cells. Somatic cell nuclear transfer (in which the nucleus of a cell from the patient is transferred into an enucleated oocyte) raises the ethical concerns of the use of human embryos and human egg donation (Rao and Condic 2008). In the iPS cell approach, adult somatic cells from the patient are reprogrammed into pluripotent cells by transduction with specific transcription factors (Takahashi and Yamanaka 2006).

R. Madonna (✉)
Institute of Cardiology, Department of Neurosciences and Imaging Laboratory of Experimental Cardiology, Center of Excellence on Aging—CESI, "G. d'Annunzio" University, Chieti Scalo, Italy
e-mail: rmadonna@unich.it

P. Ferdinandy
Department of Pharmacology and Pharmacotherapy, Semmelweis University, Budapest, Hungary

Pharmahungary Group, Szeged, Hungary

R. Schulz
Institute of Physiology, Justus-Liebig Giessen University of Giessen, Giessen, Germany

R. Madonna (ed.), *Stem Cells and Cardiac Regeneration*, Stem Cell Biology and Regenerative Medicine, DOI 10.1007/978-3-319-25427-2_3

iPS cells are a type of pluripotent stem cell artificially derived from a non-pluripotent cell—typically an adult somatic cell—by inducing a "forced" expression of specific genes. Typically, iPS cells are generated by retroviral induction of transcription factors, OCT (octamer-binding transcription factor)-4, SRY (sex determining region Y)-box 2 also known as SOX2, kruppel-like factor (KLF)-4, and c-MYC, in fibroblasts (Takahashi and Yamanaka 2006). Lentivirus and adenovirus induction, induction with other gene combinations and virus free approaches such as using plasmids, small molecules and recombinant proteins have also been reported (Takahashi and Yamanaka 2006). In addition, it has been shown that iPS cells can be generated from a variety of cell types such as pancreatic cells, meningiocytes, keratinocytes, hematopoietic cells differentiated from ESCs, and primary human hepatocytes (Takahashi and Yamanaka 2006; EMD Millipore 2008).

Beside regenerative medicine, iPS cells have several potential applications for testing mechanistic hypotheses of diseases or for drug and gene screening. They have been proposed as a new model to replace animal experiments in drug development and toxicity tests, in which animals have not proven appropriate and safe models to predict the efficacy and toxicity of drug candidates in humans.

3.2 Methods to Develop Human Induced Pluripotent Stem Cells

Many vectors of reprogramming, including adenoviruses, transposons, plasmids, ESC protein extracts, recombinant proteins, synthetic mRNAs, mature microRNAs, virus RNA, have the potential to be tested for inducing pluripotency in human somatic cells. Most methods employed so far to generate iPS cells with consistent efficiency include the transfer of foreign DNA into the target cell by retroviral vectors or non-integrating vectors. Takahashi and Yamanaka (Takahashi and Yamanaka 2006) demonstrated that the overexpression of four transcription factors, OCT (octamer-binding transcription factor)-4, SRY (sex determining region Y)-box 2 also known as SOX2, kruppel-like factor (KLF)-4, and c-MYC, can reprogram mouse somatic fibroblasts to iPS cells. These factors lead to the reactivation of the endogenous pluripotency genes encoding Oct4, Nanog, and Sox2 and to the activation of the autoregulatory loop that maintains the pluripotent state. In 2007, two research groups independently showed that reprogramming could be induced not only by OCT4, SOX2, KLF4, and c-MYC but also by alternative combinations that employ NANOG, lineage protein (LIN)-28, Estrogen-related receptor beta (ESRRB) and nuclear receptor subfamily 5 group A member (NR5A)-2 (Yu et al. 2007; Ichida et al. 2009). Research work by subsequent groups tried to overcome the issues related to the low efficiency of reprogramming. Specifically, Carey and colleagues (Carey et al. 2009) reported a tenfold more efficient generation of iPS cells by using a single polycistronic vector to transfer the reprogramming factors. Other research groups worked on the issue of cancer risk by using non-integrating adenoviral or episomal vectors (Stadtfeld et al. 2008; Kaji et al. 2009; Woltjen et al. 2009; Yu et al. 2009; Okita et al. 2010). In the last 2 years several researchers have successfully developed and implemented protein-based methods (Yu et al. 2007; Kim et al. 2009). In this approach, native and active recombinant proteins of the four reprogramming factors attached to short fusogenic peptides, such as HIV-TAT and/or to a poly-arginine domain (9–11 Arg), were generated. However, two major drawbacks of this technology are the cumbersome renaturation/refolding protocols of recombinant proteins, trapped in inclusion bodies, and the extremely low reprogramming efficiency. The presence of small molecules such as valproic acid (VPA), a histone deacetylase (HDAC) inhibitor, and Vitamin-C have been reported to improve reprogramming efficiency significantly (Esteban et al. 2010). Hence, further optimization of this technology in reprogramming somatic cells to iPS cells is needed in order to approach a similar efficiency level as obtained using viral protocols. Other reprogramming methods to generate iPS cells, beside protein transfer, are RNA-based transfer (Sul et al. 2009) "transcriptome-induced phenotype remodeling (TIPeR)". This method transfers the entire regulatory components from a target cell to a donor cell and achieves cellular reprogramming by manipulating 'whole systems' rather than a small set of master genes. Other RNA-based reprogramming strategies included the administration of synthetic modified mRNA (Warren et al. 2010) and mature double-stranded microRNAs (Miyoshi et al. 2011). MicroRNAs are short non-coding RNAs (22 base pair length), encoded by introns or intergenic regions of most of the transcribed primary RNA, that control gene expression by either inducing mRNA degradation or blocking translation (Bartel 2004). The same technology based on microRNAs can be used to direct the reprogramming of mature cells from one lineage to terminally differentiated neurogenic cells. This will be discussed later. The RNA- and transcriptome-based technology might overcome the safety concerns related with vector-based DNA transfer. A recent innovative approach to increase the efficiency of cell reprogramming has been the inactivation or inhibition of p53 by lentivirus-mediated RNA interference (RNAi) (Esteban et al. 2009; Krizhanovsky and Lowe 2009). This approach uses the random integration of shRNA cassettes by lentiviral vectors. An alternative approach to silence p53, or other genes of interest, is known as 'genome editing', initially developed in Drosophila melanogaster (Bibikova et al. 2002). The technique consists in the ability to specifically add or delete genetic information, enabling efficient and precise genetic modification via the induction of a double-strand break (DSB) in a specific genomic target sequence, followed by the generation of desired modifications during subsequent DNA break repair. The DSB is induced by a 'zinc finger nuclease' (ZFN) (Bibikova et al. 2001), which is a sequence specific endonuclease that can be customized to cleave the DNA target. Each ZFN consists of DNA-binding domain and DNA-cleaving domain which comprises the nuclease domain of FokI.

An example of genome editing to silence p53 in iPS cells is given by research work from DeKelver and colleagues (DeKelver et al. 2010). We will review later the genome editing, including the p53 silencing in the iPS cells. Whatever the technique used to silence the expression of p53, the question remains as to whether the reprogrammed iPS cells might exhibit compromised genome integrity in the absence of p53 and thus run the risk of becoming malignant.

Additional strategies in order to optimize somatic reprogramming or increase the efficiency of cardiomyocyte differentiation from iPS cells are based on manipulating intracellular levels of reactive oxygen species (ROS) or other metabolites involved in the glycolytic flux and mitochondrial respiration. Low levels of ROS in quiescent iPS cells contribute to maintaining their "pluripotency," whereas a higher level of ROS promotes differentiation, proliferation, migration, and survival of iPS cells (Chaudhari et al. 2014; Guo et al. 2013; Pereira et al. 2013). Major sources of ROS are NADPH oxidase and mitochondria (Crespo et al. 2010) (for review, see Urao and Ushio-Fukai 2013). iPS cells basically cover their energetic need through an increased glycolytic flux and less mitochondrial respiration, thereby reducing ROS formation. Indeed, stimulation of glycolysis through induction of glycolytic enzyme, and repression of mitochondrial respiration through downregulated electron transport chain complex I subunits, promotes, while blockade of glycolytic enzyme activity bluntes, reprogramming efficiency (Folmes et al. 2011; Madonna et al. 2013, 2014). Interestingly, inhibition of the mitochondrial permeability transition pore by cyclosporin A alters mitochondrial oxidative metabolism and redox signaling, which leads to differentiation of functional cardiomyocytes from iPS cells (Cho et al. 2014).

All these studies involve conversion of one somatic cell type (skin fibroblasts, keratinocytes, blood cells) to another (neuronal cells, cardiomyocytes) through the requirement for "helper" cells, such as iPS cells. Direct reprogramming of terminally differentiated cells, also called lineage reprogramming, can provide rapid production of models of human "diseases in a dish", without first passing the cells through a pluripotent state, so avoiding the challenges of time-consuming and labor-intensive iPS cell line generation.

3.3 Genome Editing and Human Induced Pluripotent Stem Cells

Constitutive or inducible transgene expression (such as the overexpression of reprogramming transcription factors) and downregulation of gene expression by RNAi or microRNA are certainly useful approach for cell reprogramming but not accurate at all, since they do not maintain the endogenous setting or function of the gene. By these approaches, the expression of an integrated transgene can be unpredictable over time due to atypical epigenetic changes occurring near its site of integration (Stewart et al. 2008). Also, the transgene can integrate in multiple site of the genome, and as it can determine disruption or activation of other genes, creating a deep change of the genetic setting of the cell. A targeted approach to genetic manipulation avoids these effects and allows for the specific editing of the genome.

Genome editing literally means "rewriting the genome". This approach allows to rewrite the genetic code of a living cell, similar to the find-and-replace function in word-processing programs. Using this approach it can be possible to design cells that build proteins not found in nature, or replace wrong proteins, apparently without disrupting the cell function. Early approaches to genome editing, dating back to 1980 with the work of Nobel Prizes Mario R. Capecchi and Oliver Smithies (Thomas et al. 1986; Nobelprize.org 2007), involved modification of genetic sequences using homologous recombination. Capecchi and Smithies developed a process for the modification of the mouse genome by modifying the DNA of mouse embryonic stem cells in culture and injecting these modified stem cells into mouse embryos. Major drawback of this approach is that modification of genomes using only homologous recombination remained a long and random process, with often low rate of homologous recombination in somatic cell types. Two alternative methods have been recently developed that enable to insert a foreign gene sequence into a live cell by triggering the cells to DNA repair processes. The first method is by site directed endonucleases (restriction enzymes), which include specific technologies such ZFNs and meganucleases. Site directed endonucleases achieve gene modification through causing DSB which triggers the cells to DNA repair mechanisms, predominantly non homologous end joining (NHEJ) as well as a low frequency of homologous recombination (HR). The second method is recombinant adeno-associated virus (rAAV) mediated genome engineering which induces high frequencies of homologous recombination alone, thus forgoing the need to perform DSB. We will review here the site directed endonuclease approach using ZFNs as this technique has been used very often for gene targeting in iPS cells, were it can increase HR efficiency (Hockemeyer et al. 2009; Zou et al. 2009). ZFNs are enzymes designed to recognize and digest (after dimerization at FokI domain) a specific DNA sequence by combining up to six C_2H_2 zinc finger proteins to 36 bp DNA template (Pabo et al. 2001). Once a DSB has been induced by ZFN, it can be repaired by intrinsic repair mechanisms of the cell, either by NHEJ repair pathway or by HR (Jackson and Bartek 2009). Compared with HR, the repair efficiency of NHEJ pathway is higher, but has the disadvantage to be more error prone because of the additions or deletions at the break point. The NHEJ pathway repair can be used to produce knockout of unwanted gene expression. To date, there are two key therapeutic applications of ZFN technology in iPS cells, that we will discuss below.

1. *Gene addition and correction*. ZFN-mediated gene addition in iPS consists of adding genes with aberrant or absent function, or mutations related with a particular disease (Lombardo et al. 2007).

   ZFN-mediated gene correction is to determine insertional mutagenesis of the disease-related gene, aimed at treating monogenic disease. By using insertional mutagenesis of lox-p flanked cassette—Cre-lox system, Zou and colleagues (2009) obtained iPS cells lines with conditional knockout of phosphatidylinositol glycan anchor biosynthesis class A (PIG-A) gene. PIG-A is mutated in hematopoietic stem cells from patients with the blood disorder paroxysmal nocturnal hemoglobinuria.

2. *Insertion into safe harbor locus*. This consists of the insertion of constitutive or inducible expressed cassettes for therapeutic expression of a particular gene, factor, drug, small molecule, or RNAi. DeKelver and colleagues (2010) demonstrated the feasibility of targeting shRNA cassettes against p53 to the AAVS1 locus in iPS cells without the use of lentivirus vectors. By this approach the authors have obtained iPS cell lines carrying p53-specific shRNA cassettes, without aberrant insertions and with long-term robust p53 gene knockdown.

   Among limitations of the genome editing, the risk of chromosomal translocation and karyotypic abnormalities have been reported in some type of cells (Brunet et al. 2009). However, several studies have already shown that iPS cells following ZFN-mediated genome targeting display no karyotypic abnormalities and changes in pluripotency (DeKelver et al. 2010).

3.4 Human Induced Pluripotent Stem Cells as Model of Human Disease

Animal models have been used so far to dissect the mechanisms and pathogenesis of diseases (Hardouin and Nagy 2000). However, genetic manipulations including over-expression, knock-down, knock-out, and knock-in strategies, are often employed in order to replicate genetic patterns linked to specific disease phenotypes (Rosenthal and Brown 2007). These manipulations make animal models not reliable for reproducing the pathophysiology and to study the nature of the disease. Furthermore, because the genetic differences between human and animals may cause diverse phenotypes, a number of diseases cannot be properly simulated with animals (Rosenthal and Brown 2007). With iPS cell technology, it is possible to produce disease-specific stem cells that carry the genome of the donor and mimic human disease (Colman 2008). This new approach consists into screen patients for a genetic cause of disease (genetic mutation), develop cell lines, take them back to iPS cells, differentiate them into one or more cell types that develop the disease. It is important to remark that the choice of the donor cell to generate iPS cells must be dictated by the type of the disease that one want to model in vitro. This is because of the existence of epigenetic memory that contributes to the phenotypic fate of iPS cells. These cell lines are then expected to develop a phenotype or a signal that can be used to study disease mechanism. The ability to dedifferentiate patient-specific cells back to stem cells and to redifferentiate these cells into specific lineages allows to reproduce human diseases in a "petri dish" and to explore their progression in different tissues (Callaway 2011). However, while this is true, there are several potential drawbacks to the use of iPS cells, including the potential for karyotypic abnormalities to emerge in these cells with prolonged culture, and the risk of failure to erase memory of the tissue of origin of the adult cell from which iPS cells were derived. These aspects will be discussed later.

Approximately 10 % of patients with amyotrophic lateral sclerosis, 7 % with early-onset alzheimer's disease, and less than 1 % with Parkinson's disease are believed to have familial variants, and only a subset of these patients carry known mutations (Koch et al. 2009). Recently, the generation of patient-specific iPS cells has been successfully established from patients with amyotrophic lateral sclerosis, spinal muscular atrophy, Parkinson disease, familial dysautonomia, Fanconi anemia, adenosine deaminase deficiency, Schwachman-Bodian-Diamond syndrome, Gaucher disease, Duchenne and Becker muscular dystrophy, Huntington disease, type 1 diabetes mellitus, Down syndrome and Lesch-Nyhan syndrome (Dimos et al. 2008; Park et al. 2008; Wernig et al. 2008; Ebert et al. 2009; Lee et al. 2009; Raya et al. 2009; Soldner et al. 2009).

The spinal muscular atrophy (SMA) includes severe infantile Werdnig-Hoffmann disease (at early onset between 0 and 6 month of age) and the mild Kugelberg-Welander disease (at later onset after 18 months of age). These syndromes are associated with mutation of the gene encoding for Survival of Motor Neuron (SMN) protein (Prior 2007). Humans have the two copies of the gene—known as SMN1 and SMN2—that differ by only a few base pairs. The SMN2 gene is due to a silent mutation that occurs at the splice junction of intron 6 to exon 7. This mutation affects splicing of the SMN2 pre-RNA, that results in about 90 % of the transcripts being inappropriately spliced into a form that excludes exon 7. This shorter mRNA transcript codes for a shorter SMN protein, which is rapidly degraded (Prior 2007). SMA is caused by loss of the SMN1 gene from both chromosomes, and is inherited as autosomal recessive disease, in which the defective gene is located on an autosome, and two copies of the gene—one from each parent—are required to inherit the disorder. iPS cells generated from adult cells that were isolated from patients have been shown to differentiate into motor neurons in vitro, but with less number and smaller size compared with normal iPS cell-derived counterparts. Accordingly with the phenotype of the disease, the SMN1 gene in these motor neurons is less expressed, and the formation of synapse is delayed (Ebert et al. 2009).

Further example of autosomal recessive disease is Familial Dysautonomia (FD), also known as Riley–Day syndrome, a fatal degenerative disease of the autonomic nervous system which affects the development and survival of sensory, sympathetic and some parasympathetic neurons in the autonomic and sensory nervous system (Axelrod 2004). FD is the result of mutations in the intron 20 of the gene on chromosome nine encoding for the IKAP (IkB kinase complex associated protein). The most common mutation is a conversion of T>C, resulting in shift splicing that generates an IKAP transcript lacking exon 20. Translation of this mRNA results in a truncated and non-functional IKAP protein, which determines the disease (Lee et al. 2009). In the iPS cells derived from these patients, there is a co-existence of both correct and incorrect splicing pattern in the iPS cell-derived neuron cell lines (Lee et al. 2009). In this study, iPS cells have been successfully used to test if the plant hormone kinetin can reduce the proportion of incorrect splicing (Lee et al. 2009).

The establishment of cardiac disease-specific iPS cells from patients with congenital long-QT syndrome (LQTS), Brugada syndrome, catecholaminergic polymorphic ventricular tachycardia, arrhythmogenic right ventricular cardiomyopathy, and other genetic arrhythmias, is advisable since it would shed some light on these disease mechanisms, resulting in the discovery of new therapies. In the LQTS, patients are characterized by a reduction of outward potassium current (IKs, IKr, or IK1) or increase of inward current (INa or ICa), which lead to delayed repolarization of the heart. This results in a prolonged QT interval and increased risk of episodes of torsade de pointes (TDP, a form of irregular heartbeat that originates from the ventricles) (Abrams et al. 2010). Genetic studies have identified five forms of LQTS caused by mutations in ion channel genes located on chromosomes 3(LQT3), 7(LQT2), 11(LQT1), and 21(LQT5 and LQT6) (Hedley et al. 2009). Recording each ion channel current of the ventricular cardiomyocytes derived from patient-specific iPS cells is expected to reveal diagnostic abnormalities of these currents. iPS cells would also help understand pathological mechanisms of such currents and develop appropriately personalized therapy for each patient by screening drugs capable to correct current abnormalities. Moretti and colleagues (Moretti et al. 2010) have recently generated patient-specific pluripotent stem cells from members of a family affected by LQTS type 1, an autosomal dominant missense mutation in the KCNQ1 gene that encodes the repolarizing potassium channel mediating the delayed rectifier IKs current. Authors harvested dermal fibroblasts from two family members and two healthy controls, and infected them with retroviral vectors encoding the human transcription factors OCT3/4, SOX2, KLF4, and c-MYC to generate pluripotent stem cells. In this study, authors showed the capacity of the cells from patients with LQTS type 1 to differentiate into contractile cardiomyocytes by the embryoid-body-differentiation system, to exhibit prolongation of the action potential, altered IKs activation and deactivation properties, and an abnormal response to catecholamine stimulation, with a protective effect of beta-blockade. These data thus show that iPS cells can model and recapitulate aspects of genetic cardiac diseases. Further research work from Itzhaki and colleagues (Itzhaki et al. 2011) showed the ability of human iPS cells to model the abnormal functional phenotype of LQTS. Here, cultured fibroblasts have been harvested from family members affected by LQTS type 2, an autosomal dominant missense mutation in the KCNQ2 gene that affect the pore-forming region of the KCNH2 (also known as HERG)-encoded potassium channel (Nakajima et al. 1998). Fibroblasts were then reprogrammed to generate LQTS patient-specific human iPS clones after transduction with retroviral vectors encoding for SOX2, KLF4 and OCT4, followed by valproic acid treatment (Huangfu et al. 2008). Authors showed significant prolongation of the action potential duration in LQTS human iPS cell-derived cardiomyocytes when compared to healthy control cells, due to significant reduction of the rectifier cardiac potassium current IK. Importantly, LQTS-derived cells also showed marked arrhythmogenicity, characterized by early-after depolarizations and triggered arrhythmias.

The iPS cell-derived cardiomyocytes have also recently been used to further understand the mechanisms underlying cardiac disease phenotype due to disorders of the RAS/MAPK signaling pathway. The ‘RASopathies’ are a family of autosomal dominant developmental disorders caused by mutations in the RAS/MAPK pathway, which generally result in increased signaling. Examples of RASopathies are Noonan syndrome and its related disorder LEOPARD syndrome (Bentires-Alj et al. 2006). Noonan syndrome (NS), named after Dr. Jacqueline Noonan, is an autosomal dominant congenital disorder affecting approximately 1 in 1000 children worldwide. Principal features of NS include congenital heart defect (50 % pulmonary valve stenosis, 20 % hypertrophic cardiomyopathy and 10 % septal defects), short stature, learning problems, pectus excavatum, impaired blood clotting and facial abnormalities (Curcic-Stojkovic et al. 1978). LEOPARD syndrome, also known as Cardiocutaneous syndrome, is a rare autosomal dominant, multisystem disease characterized by the following seven conditions, the first letters of which spell LEOPARD: Lentigines, Electrocardiographic conduction abnormalities, Ocular hypertelorism, Pulmonary stenosis, Abnormal genitalia, Retarded growth, Deafness (Coppin and Temple 1997). Pulmonic stenosis and hypertrophic cardiomyopathy are seen in 23 % and 80 % of patients with LEOPARD syndrome, respectively. LEOPARD syndrome is caused by a missense mutation in the protein tyrosine phosphatase, non-receptor type 11 gene (PTPN11), which encodes the Src homology-2 (SH2) domain-containing nontransmembrane protein tyrosine phosphatase SHP2. The mutation results in mixed gain-of-function and dominant negative effects, while a different missense mutation of the same gene that causes NS, results in gain-of-function phenotype (Bentires-Alj et al. 2006). Carvajal-Vergara and colleagues (2010) harvested dermal fibroblasts from two LEOPARD syndrome patients, and infected them with retroviral vectors encoding the human transcription factors OCT3/4, SOX2, KLF4, and c-MYC to generate iPS cells. The authors showed the capability of these iPS cells to differentiate into cardiomyocytes modeling the feature of hypertrophic cardiomyopathy. When compared with cardiomyocytes derived from wild-type iPS cells generated from healthy sibling control, LEOPARD-derived cardiomyocytes had increased cell size, higher degree of sarcomeric organization and preferential localization of NFATC4 transcription factor in the nucleus. iPS cells generated from LEOPARD patients displayed also decreased ability to respond to external growth factors, such as basic fibroblast growth factors, and decreased phosphorylation of ERK1/2, indicating the presence of perturbed RAS-MAPK signal transduction as mechanism promoting the disease phenotype. The possibility to use an in vitro system like iPS to screen patients for the presence of epigenetic or functional alteration of these candidate genes might help prevent or treat cardiovascular complications in diabetic patients. iPS cells can enable a comparison to be made of the pathogenesis of a particular disease among patients and can also offer opportunity for drug screening and toxicity testing. An example is given by cardiac drug screening, for which human cardiomyocytes are not practicable (Sartipy et al. 2007). Tanaka and colleagues gave proof of this concept (Tanaka et al. 2009). Authors showed that iPS cell-derived cardiomyocytes can represent a promising in vitro model of cardiac electrophysiologic studies and drug screening, since it is possible to derive functional cardiomyocytes from iPS cells with similar expression pattern and contractile function compared with normal counterparts (Tanaka et al. 2009). In this study, human iPS cells were driven to differentiate into functional cardiomyocytes, which expressed cardiac markers including Nkx2.5, GATA4, and atrial natriuretic peptide. The iPS cell-derived cardiomyocytes were analyzed using a multi electrode assay. The application of ion channel inhibitors resulted in dose-dependent changes to the field potential waveform, and these changes were identical to those induced in the native cardiomyocytes (Tanaka et al. 2009). Finally, iPS cells can allow the development of personalized therapies using a patient's own cells, by re-engineering the cells to correct disease-causing defects before reintroducing them into the body (Somers et al. 2010; Yoshizaki et al. 2010).

References

Abrams DJ, Perkin MA, Skinner JR (2010) [Long QT syndrome]. Praxis (Bern 1994) 99:854–858

Axelrod FB (2004) Familial dysautonomia. Muscle Nerve 29:352–363

Bartel DP (2004) MicroRNAs: genomics, biogenesis, mechanism, and function. Cell 116:281–297

Bentires-Alj M, Kontaridis MI, Neel BG (2006) Stops along the RAS pathway in human genetic disease. Nat Med 12:283–285

Bibikova M, Carroll D, Segal DJ, Trautman JK, Smith J, Kim YG, Chandrasegaran S (2001) Stimulation of homologous recombination through targeted cleavage by chimeric nucleases. Mol Cell Biol 21:289–297

Bibikova M, Golic M, Golic KG, Carroll D (2002) Targeted chromosomal cleavage and mutagenesis in Drosophila using zinc-finger nucleases. Genetics 161:1169–1175

Brunet E, Simsek D, Tomishima M, DeKelver R, Choi VM, Gregory P, Urnov F, Weinstock DM, Jasin M (2009) Chromosomal translocations induced at specified loci in human stem cells. Proc Natl Acad Sci U S A 106:10620–10625

Callaway E (2011) Cells snag top modelling job. Nature 469:279

Carey BW, Markoulaki S, Hanna J, Saha K, Gao Q, Mitalipova M, Jaenisch R (2009) Reprogramming of murine and human somatic cells using a single polycistronic vector. Proc Natl Acad Sci U S A 106:157–162

Carvajal-Vergara X, Sevilla A, D'Souza SL, Ang YS, Schaniel C, Lee DF, Yang L, Kaplan AD, Adler ED, Rozov R, Ge Y, Cohen N, Edelmann LJ, Chang B, Waghray A, Su J, Pardo S, Lichtenbelt KD, Tartaglia M, Gelb BD, Lemischka IR (2010) Patient-specific induced pluripotent stem-cell-derived models of LEOPARD syndrome. Nature 465:808–812

Chaudhari P, Ye Z, Jang YY (2014) Roles of reactive oxygen species in the fate of stem cells. Antioxid Redox Signal 20(12):1881–1890

Cho SW, Park JS, Heo HJ, Park SW, Song S, Kim I, Han YM, Yamashita JK, Youm JB, Han J, Koh GY (2014) Dual modulation of the mitochondrial permeability transition pore and redox signaling synergistically promotes cardiomyocyte differentiation from pluripotent stem cells. J Am Heart Assoc 3(2), e000693

Colman A (2008) Induced pluripotent stem cells and human disease. Cell Stem Cell 3:236–237

Coppin BD, Temple IK (1997) Multiple lentigines syndrome (LEOPARD syndrome or progressive cardiomyopathic lentiginosis). J Med Genet 34:582–586

Crespo FL, Sobrado VR, Gomez L, Cervera AM, McCreath KJ (2010) Mitochondrial reactive oxygen species mediate cardiomyocyte formation from embryonic stem cells in high glucose. Stem Cells 28(7):1132–1142

Curcic-Stojkovic O, Nikolic L, Obradovic D, Krstic A, Radic A (1978) Noonan's syndrome. (Male Turner's syndrome, Turner-like syndrome). Med Pregl 31:299–303

DeKelver RC, Choi VM, Moehle EA, Paschon DE, Hockemeyer D, Meijsing SH, Sancak Y, Cui X, Steine EJ, Miller JC, Tam P, Bartsevich VV, Meng X, Rupniewski I, Gopalan SM, Sun HC, Pitz KJ, Rock JM, Zhang L, Davis GD, Rebar EJ, Cheeseman IM, Yamamoto KR, Sabatini

DM, Jaenisch R, Gregory PD, Urnov FD (2010) Functional genomics, proteomics, and regulatory DNA analysis in isogenic settings using zinc finger nuclease-driven transgenesis into a safe harbor locus in the human genome. Genome Res 20:1133–1142

Dimos JT, Rodolfa KT, Niakan KK, Weisenthal LM, Mitsumoto H, Chung W, Croft GF, Saphier G, Leibel R, Goland R, Wichterle H, Henderson CE, Eggan K (2008) Induced pluripotent stem cells generated from patients with ALS can be differentiated into motor neurons. Science 321:1218–1221

Ebert AD, Yu J, Rose FF Jr, Mattis VB, Lorson CL, Thomson JA, Svendsen CN (2009) Induced pluripotent stem cells from a spinal muscular atrophy patient. Nature 457:277–280

Esteban MA, Xu J, Yang J, Peng M, Qin D, Li W, Jiang Z, Chen J, Deng K, Zhong M, Cai J, Lai L, Pei D (2009) Generation of induced pluripotent stem cell lines from Tibetan miniature pig. J Biol Chem 284:17634–17640

Esteban MA, Wang T, Qin B, Yang J, Qin D, Cai J, Li W, Weng Z, Chen J, Ni S, Chen K, Li Y, Liu X, Xu J, Zhang S, Li F, He W, Labuda K, Song Y, Peterbauer A, Wolbank S, Redl H, Zhong M, Cai D, Zeng L, Pei D (2010) Vitamin C enhances the generation of mouse and human induced pluripotent stem cells. Cell Stem Cell 6:71–79

Folmes CD, Nelson TJ, Martinez-Fernandez A, Arrell DK, Lindor JZ, Dzeja PP, Ikeda Y, Perez-Terzic C, Terzic A (2011) Somatic oxidative bioenergetics transitions into pluripotency-dependent glycolysis to facilitate nuclear reprogramming. Cell Metab 14(2):264–271

Guo CW, Kawakatsu M, Idemitsu M, Urata Y, Goto S, Ono Y, Hamano K, Li TS (2013) Culture under low physiological oxygen conditions improves the stemness and quality of induced pluripotent stem cells. J Cell Physiol 228(11):2159–2166

Hardouin SN, Nagy A (2000) Mouse models for human disease. Clin Genet 57:237–244

Hedley PL, Jorgensen P, Schlamowitz S, Wangari R, Moolman-Smook J, Brink PA, Kanters JK, Corfield VA, Christiansen M (2009) The genetic basis of long QT and short QT syndromes: a mutation update. Hum Mutat 30:1486–1511

Hockemeyer D, Soldner F, Beard C, Gao Q, Mitalipova M, DeKelver RC, Katibah GE, Amora R, Boydston EA, Zeitler B, Meng X, Miller JC, Zhang L, Rebar EJ, Gregory PD, Urnov FD, Jaenisch R (2009) Efficient targeting of expressed and silent genes in human ESCs and iPSCs using zinc-finger nucleases. Nat Biotechnol 27:851–857

Huangfu D, Maehr R, Guo W, Eijkelenboom A, Snitow M, Chen AE, Melton DA (2008) Induction of pluripotent stem cells by defined factors is greatly improved by small-molecule compounds. Nat Biotechnol 26:795–797

Ichida JK, Blanchard J, Lam K, Son EY, Chung JE, Egli D, Loh KM, Carter AC, Di Giorgio FP, Koszka K, Huangfu D, Akutsu H, Liu DR, Rubin LL, Eggan K (2009) A small-molecule inhibitor of tgf-Beta signaling replaces sox2 in reprogramming by inducing nanog. Cell Stem Cell 5:491–503

Itzhaki I, Maizels L, Huber I, Zwi-Dantsis L, Caspi O, Winterstern A, Feldman O, Gepstein A, Arbel G, Hammerman H, Boulos M, Gepstein L (2011) Modelling the long QT syndrome with induced pluripotent stem cells. Nature 471:225–229

Jackson SP, Bartek J (2009) The DNA-damage response in human biology and disease. Nature 461:1071–1078

Kaji K, Norrby K, Paca A, Mileikovsky M, Mohseni P, Woltjen K (2009) Virus-free induction of pluripotency and subsequent excision of reprogramming factors. Nature 458:771–775

Kim JB, Greber B, Arauzo-Bravo MJ, Meyer J, Park KI, Zaehres H, Scholer HR (2009) Direct reprogramming of human neural stem cells by OCT4. Nature 461:649–653

Koch P, Kokaia Z, Lindvall O, Brustle O (2009) Emerging concepts in neural stem cell research: autologous repair and cell-based disease modelling. Lancet Neurol 8:819–829

Krizhanovsky V, Lowe SW (2009) Stem cells: the promises and perils of p53. Nature 460:1085–1086

Lee G, Papapetrou EP, Kim H, Chambers SM, Tomishima MJ, Fasano CA, Ganat YM, Menon J, Shimizu F, Viale A, Tabar V, Sadelain M, Studer L (2009) Modelling pathogenesis and treatment of familial dysautonomia using patient-specific iPSCs. Nature 461:402–406

Lombardo A, Genovese P, Beausejour CM, Colleoni S, Lee YL, Kim KA, Ando D, Urnov FD, Galli C, Gregory PD, Holmes MC, Naldini L (2007) Gene editing in human stem cells using

zinc finger nucleases and integrase-defective lentiviral vector delivery. Nat Biotechnol 25:1298–1306
Madonna R, Gorbe A, Ferdinandy P, De Caterina R (2013) Glucose metabolism, hyperosmotic stress, and reprogramming of somatic cells. Mol Biotechnol 55(2):169–178
Madonna R, Geng YJ, Shelat H, Ferdinandy P, De Caterina R (2014) High glucose-induced hyperosmolarity impacts proliferation, cytoskeleton remodeling and migration of human induced pluripotent stem cells via aquaporin-1. Biochim Biophys Acta 1842(11):2266–2275
EMD Millipore (2008) ENStem-A adherent human neural progenitors: a new source of primary human neural cells. Application Note
Miyoshi N, Ishii H, Nagano H, Haraguchi N, Dewi DL, Kano Y, Nishikawa S, Tanemura M, Mimori K, Tanaka F, Saito T, Nishimura J, Takemasa I, Mizushima T, Ikeda M, Yamamoto H, Sekimoto M, Doki Y, Mori M (2011) Reprogramming of mouse and human cells to pluripotency using mature microRNAs. Cell Stem Cell 8:633–638
Moretti A, Bellin M, Welling A, Jung CB, Lam JT, Bott-Flugel L, Dorn T, Goedel A, Hohnke C, Hofmann F, Seyfarth M, Sinnecker D, Schomig A, Laugwitz KL (2010) Patient-specific induced pluripotent stem-cell models for long-QT syndrome. N Engl J Med 363:1397–1409
Nakajima T, Furukawa T, Tanaka T, Katayama Y, Nagai R, Nakamura Y, Hiraoka M (1998) Novel mechanism of HERG current suppression in LQT2: shift in voltage dependence of HERG inactivation. Circ Res 83:415–422
Nobelprize.org (2007) The Nobel prize in physiology or medicine 2007
Okita K, Hong H, Takahashi K, Yamanaka S (2010) Generation of mouse-induced pluripotent stem cells with plasmid vectors. Nat Protoc 5:418–428
Pabo CO, Peisach E, Grant RA (2001) Design and selection of novel Cys2His2 zinc finger proteins. Annu Rev Biochem 70:313–340
Park IH, Arora N, Huo H, Maherali N, Ahfeldt T, Shimamura A, Lensch MW, Cowan C, Hochedlinger K, Daley GQ (2008) Disease-specific induced pluripotent stem cells. Cell 134:877–886
Pereira SL, Grãos M, Rodrigues AS, Anjo SI, Carvalho RA, Oliveira PJ, Arenas E, Ramalho-Santos J (2013) Inhibition of mitochondrial complex III blocks neuronal differentiation and maintains embryonic stem cell pluripotency. PLoS One 8(12):e82095
Prior TW (2007) Spinal muscular atrophy diagnostics. J Child Neurol 22:952–956
Rao M, Condic ML (2008) Alternative sources of pluripotent stem cells: scientific solutions to an ethical dilemma. Stem Cells Dev 17:1–10
Raya A, Rodriguez-Piza I, Guenechea G, Vassena R, Navarro S, Barrero MJ, Consiglio A, Castella M, Rio P, Sleep E, Gonzalez F, Tiscornia G, Garreta E, Aasen T, Veiga A, Verma IM, Surralles J, Bueren J, Izpisua Belmonte JC (2009) Disease-corrected haematopoietic progenitors from Fanconi anaemia induced pluripotent stem cells. Nature 460:53–59
Rosenthal N, Brown S (2007) The mouse ascending: perspectives for human-disease models. Nat Cell Biol 9:993–999
Sartipy P, Bjorquist P, Strehl R, Hyllner J (2007) The application of human embryonic stem cell technologies to drug discovery. Drug Discov Today 12:688–699
Soldner F, Hockemeyer D, Beard C, Gao Q, Bell GW, Cook EG, Hargus G, Blak A, Cooper O, Mitalipova M, Isacson O, Jaenisch R (2009) Parkinson's disease patient-derived induced pluripotent stem cells free of viral reprogramming factors. Cell 136:964–977
Somers A, Jean JC, Sommer CA, Omari A, Ford CC, Mills JA, Ying L, Sommer AG, Jean JM, Smith BW, Lafyatis R, Demierre MF, Weiss DJ, French DL, Gadue P, Murphy GJ, Mostoslavsky G, Kotton DN (2010) Generation of transgene-free lung disease-specific human induced pluripotent stem cells using a single excisable lentiviral stem cell cassette. Stem Cells 28:1728–1740
Stadtfeld M, Nagaya M, Utikal J, Weir G, Hochedlinger K (2008) Induced pluripotent stem cells generated without viral integration. Science 322:945–949
Stewart R, Yang C, Anyfantis G, Przyborski S, Hole N, Strachan T, Stojkovic M, Keith WN, Armstrong L, Lako M (2008) Silencing of the expression of pluripotent driven-reporter genes stably transfected into human pluripotent cells. Regen Med 3:505–522

Sul JY, Wu CW, Zeng F, Jochems J, Lee MT, Kim TK, Peritz T, Buckley P, Cappelleri DJ, Maronski M, Kim M, Kumar V, Meaney D, Kim J, Eberwine J (2009) Transcriptome transfer produces a predictable cellular phenotype. Proc Natl Acad Sci U S A 106:7624–7629

Takahashi K, Yamanaka S (2006) Induction of pluripotent stem cells from mouse embryonic and adult fibroblast cultures by defined factors. Cell 126:663–676

Tanaka T, Tohyama S, Murata M, Nomura F, Kaneko T, Chen H, Hattori F, Egashira T, Seki T, Ohno Y, Koshimizu U, Yuasa S, Ogawa S, Yamanaka S, Yasuda K, Fukuda K (2009) In vitro pharmacologic testing using human induced pluripotent stem cell-derived cardiomyocytes. Biochem Biophys Res Commun 385:497–502

Thomas KR, Folger KR, Capecchi MR (1986) High frequency targeting of genes to specific sites in the mammalian genome. Cell 44:419–428

Urao N, Ushio-Fukai M (2013) Redox regulation of stem/progenitor cells and bone marrow niche. Free Radic Biol Med 54:26–39

Warren L, Manos PD, Ahfeldt T, Loh YH, Li H, Lau F, Ebina W, Mandal PK, Smith ZD, Meissner A, Daley GQ, Brack AS, Collins JJ, Cowan C, Schlaeger TM, Rossi DJ (2010) Highly efficient reprogramming to pluripotency and directed differentiation of human cells with synthetic modified mRNA. Cell Stem Cell 7:618–630

Wernig M, Zhao JP, Pruszak J, Hedlund E, Fu D, Soldner F, Broccoli V, Constantine-Paton M, Isacson O, Jaenisch R (2008) Neurons derived from reprogrammed fibroblasts functionally integrate into the fetal brain and improve symptoms of rats with Parkinson's disease. Proc Natl Acad Sci U S A 105:5856–5861

Woltjen K, Michael IP, Mohseni P, Desai R, Mileikovsky M, Hamalainen R, Cowling R, Wang W, Liu P, Gertsenstein M, Kaji K, Sung HK, Nagy A (2009) piggyBac transposition reprograms fibroblasts to induced pluripotent stem cells. Nature 458:766–770

Yoshizaki S, Nishi M, Kondo A, Kojima Y, Yamamoto N, Ryo A (2010) Vaccination with human induced pluripotent stem cells creates an antigen-specific immune response against HIV-1 gp160. Front Microbiol 2:27

Yu J, Vodyanik MA, Smuga-Otto K, Antosiewicz-Bourget J, Frane JL, Tian S, Nie J, Jonsdottir GA, Ruotti V, Stewart R, Slukvin II, Thomson JA (2007) Induced pluripotent stem cell lines derived from human somatic cells. Science 318:1917–1920

Yu J, Hu K, Smuga-Otto K, Tian S, Stewart R, Slukvin II, Thomson JA (2009) Human induced pluripotent stem cells free of vector and transgene sequences. Science 324:797–801

Zou J, Maeder ML, Mali P, Pruett-Miller SM, Thibodeau-Beganny S, Chou BK, Chen G, Ye Z, Park IH, Daley GQ, Porteus MH, Joung JK, Cheng L (2009) Gene targeting of a disease-related gene in human induced pluripotent stem and embryonic stem cells. Cell Stem Cell 5:97–110

Chapter 4
Tissue Specific Progenitors/Stem Cells for Cardiac Regeneration

Rosalinda Madonna

4.1 Bone Marrow-Derived Stem Cells

Bone marrow-derived stem cells are currently the most commonly used cells in cell transplantation therapy. Strauer et al. (2002) treated 10 patients with acute ST elevation myocardial infarcts using autologous mononuclear bone marrow cells through a balloon catheter placed into an infarct-related coronary artery (IRA) following percutaneous transluminal coronary angioplasty (PTCA). Three months later, the infarct region in the BMC-treated group was decreased compared with that at baseline (from 30 ± 13 to 12 ± 7 %, $p = 0.005$); segmental wall movement velocity increased significantly only in the BMC-treated group (from 2.0 ± 1.1 to 4.0 ± 2.6 cm/s, $p = 0.028$). Ejection fraction increased in both groups, albeit not significantly. Dobutamine stress echocardiography, radionuclide ventriculography, and right heart catheterization of the BMC-treated group showed improved stroke volume index, left ventricular end diastolic volume, and contractility. These results demonstrated that intracoronary injection of autologous mononuclear BMCs is safe and apparently effective under clinical conditions. The TOPCARE-AMI trial led by Zeiher et al. included 20 patients with acute myocardial infarction (AMI) who received intracoronary infusion of bone marrow derived progenitor cells ($n = 9$) and blood-derived progenitor cells ($n = 11$) (Assmus et al. 2002). This treatment resulted in significantly increased global LVEF, improved regional wall motion in the infarct zone, and profoundly reduced end-systolic left ventricular volumes at 4 months follow-up. Compared with a non-randomized matched control group, LVEF only slightly increased and end-systolic volume remained unchanged. These preliminary data demonstrated the feasibility, safety, and

R. Madonna (✉)
Institute of Cardiology, Department of Neurosciences and Imaging Laboratory of Experimental Cardiology, Center of Excellence on Aging—CESI, "G. d'Annunzio" University, Chieti Scalo, Italy
e-mail: rmadonna@unich.it

R. Madonna (ed.), *Stem Cells and Cardiac Regeneration*, Stem Cell Biology and Regenerative Medicine, DOI 10.1007/978-3-319-25427-2_4

potentially beneficial effects of intracoronary autologous progenitor cells in AMI patients. Another study from the same trial included 28 patients who received circulating blood- or bone marrow-derived progenitor cells (Britten et al. 2003). The results demonstrated that progenitor cell transplantation played an important role in improving LVEF and ameliorating ventricular remodeling after infarction. Although these two studies did not reveal the cellular mechanisms associated with the improvement of LVEF after cell therapy, they did provide preliminary evidence that autologous progenitor cells can migrate into damaged myocardium and exert beneficial effects on heart function. Fuchs et al. (2003) injected bone marrow cells in 10 patients with end-stage heart failure by a strategy of endocardial delivery that utilized left ventricular electromechanical guidance. The mean time for bone marrow aspiration and processing was 2.5 h, and among the CD45/CD34 double positive cells injected, 85 ± 14 % co-expressed stem cell factor receptor CD117. Three months later, there were no reports of arrhythmia, infection, myocardial inflammation, or fibrosis. Although there was no change in EF, 8 of the 10 patients had less severe angina symptoms, and 9 patients showed prolonged exercise duration times on a treadmill test. Despite not showing efficacy, this study did demonstrate the possibility of transendocardial administration of freshly aspirated autologous bone marrow cells. Perin et al. (2003) have demonstrated the safety and efficacy of transendocardial injection of autologous bone marrow mononuclear cells in patients with end-stage ischemic heart disease. Fourteen patients were successfully injected with such cells. At 2 months follow-up, their symptoms of heart failure and angina had improved, when compared with those in the control group. METs and VO2-max on a treadmill test also improved in the cell-treated group. LVEF by echocardiography increased in both the treated and control groups. However, 4 months later, after repeat left ventriculography and electromechanical mapping were performed in the cell-treated group, the results showed increased LVEF (from 20 to 29 %, $p = 0.0003$) and improved contraction at the injection site. Also, in an experimental ischemic canine model, bone marrow-derived mesenchymal stem cells (BMM-SCs) have been shown to differentiate into endothelial cells and enhance neovascularization after myocardial injection via the NOGA system (Perin et al. 2002). The results of these studies indicate that bone marrow-derived stem cells can beneficially increase perfusion in ischemic myocardium. In addition, transendocardial injection of bone marrow cells is a safe and effective treatment option for end-stage heart failure patients.

4.2 Endothelial Progenitor Cells

Endothelial progenitor cells (EPCs) can be identified in adult peripheral blood, bone marrow, and human umbilical cord blood. Because EPCs have in common with hematopoietic stem cells certain cell surface antigens, they are considered to derive from a common precursor (Kalka et al. 2000). Current experiments suggest that EPCs play an important role in vasculogenesis by differentiating into vascular endothelial cells, inhibiting ventricular remodeling through improvement in myocardial blood supply instead of increasing the number of contracting muscle cells as skeletal myoblasts or ES cells do (Kamihata et al. 2001). Combination therapy with cardiogenesis and vasculogenesis by cell transplantation is now being considered as a novel promising strategy for repairing the ischemic damaged heart.

4.3 Adipose Tissue-Derived Stem Cells

There are major limitations in the use of adult stem cells harvested from essential organs such as muscle, skin, brain. liver and bone marrow. The pain and expense associated with the procedures, and the risk of donor site tissue morbidity are significant concerns. For the bone marrow, traditional tissue procurement procedure frequently requires general or spinal anesthesia and may yield low numbers of stem cells upon processing (approximately 1 bone marrow-derived stem cell per 10^5 adherent stromal cells) (Pittenger et al. 1999). From a practical standpoint, low stem cell numbers necessitate an ex vivo expansion step to obtain clinically significant cell numbers. Such a step is time consuming, expensive, and risks cell contamination and loss. Adipose tissue may represent an ideal source of autologous stem cells since easy to obtain by collagene digestion, results in minimal patient discomfort, yet capable of yielding cell number substantial enough to obviate extensive expansion in culture. Humans have abundant subcutaneous or abdominal fat deposits, and adipose tissue can be easily removed by aspiration from these locations. The adipose tissue, like the bone marrow, is derived from the embryonic mesoderm and contains a heterogeneous stromal cell population, including mesenchymal stem cells (Zuk et al. 2001) and endothelial cell progenitors (Miranville et al. 2004). Injection of ADSCs has been recently shown to improve neovascularisation in the ischemic hindlimb (Miranville et al. 2004) and the formation of osteoid matrix in immune-tolerant mice (Hicok et al. 2004). In the same way, ADSCs increased the functional capacity of damaged skeletal muscle in vivo (Bacou et al. 2004).

ADSCs have been shown to exhibit in vitro differentiation into the cardiomyocyte lineage (Rangappa et al. 2003). In these reports, different approaches have been used. Rangappa et al. described first the appearance of beating cells upon treatment with 5-azacytidine (Rangappa et al. 2003). They treated twice-passaged ADSCs with 5-azacytidine at 1, 3, 6, and 9 μmol/L and incubated for 12, 24, 48 or 72 h. After the respective incubation periods, the cells were washed and replaced with fresh RPMI medium (Gibco, St. Louis CA). The cells were observed daily, and the medium changed once every 3 days, until the experiment was terminated at 2 months. At 1 week, after treatment with 5-azacytidine, the cells began to change their morphology, by showing multinucleation. At 2 weeks, 30 % of the cells aggregated and formed a ball-like appearance. At 3 weeks, cell aggregates began to beat spontaneously. Immunostaining against cardiac-specific markers showed strong positivity for myosin heavy chain, α-sarcomeric actinin and troponin I in the beating cell aggregates. Planat-Benard and colleagues described spontaneous cardiomyocyte differentiation of ADSCs from primary culture, without any chemical treatment (Planat-Benard et al. 2004). In their protocol, isolated stromal vascular fraction (SVF) cells from the adipose tissue were directly plated in semisolid methylcellulose medium (MethoCult, Stem Cell Technologies, Vancouver CA) without previous cell expansion or culture selection. Beginning at 6 days of culture, the emergence of clones with rounded cells and small tube cells was identified. From these clones, some rounded cells independently started a contractile activity at days 11–14 after plating. After 20–30 days, cohesive groups of cells appeared, with the presence of branching fibers and sharing tight connections, all beating at a single rate. The percentage of beating clones counted at 20 days were from 0.02 to 0.07 % of plated SVF cells. The cardiomyocyte phenotype was confirmed by assessing the expression of specific cardiac markers, immunocytochemistry staining and ultrastructural analysis, revealing the presence of ventricular and atrial type of cells. Electrophysiological studies revealed a pacemaker activity of the cells. Functional studies showed that adrenergic agonists stimulated the beating rate, whereas cholinergic agonist decreased it.

Culture-expanded cells in control medium, isolated from the bone marrow (bone marrow-derived stromal cells, BMSCs), and from the adipose tissue (ADSCs) have been shown to give rise to apparently homogeneous populations with similar cell size and cell surface markers (Morizonoc et al. 2003). For the analysis at flow cytometry, primary cultures of BMSCs and ADSCs were expanded until passage 4 in control medium (Dulbecco's modified Eagle's medium, 10 % fetal bovine serum, 1 % antibiotic-antimycotic, all from Gibco) (Morizonoc et al. 2003). Both populations have been shown to express CD13, CD29 (β1-integrin), CD44, CD58, CD90, CD105 (endoglin) and CD166. SH-3, which recognizes an epitope present on CD73 (ecto-5′-nucleotidase), is uniformly positive for both ADSCs and BMSCs, and may act to mediate cell-cell interaction (Barry et al. 2001). STRO-1, a marker for cells with multi-lineage potential, is also expressed, albeit at low levels, by both ADSCs and BMSCs (Simmons and Torok-Storb 1991; Gronthos et al. 1994). Hovewer, certain key differences in surface marker expression between marrow and adipose-derived cells are present. Differences are in the expression of adhesion molecules with know function in the homing and mobilization of hematopoietic stem cells. For example, ADSCs express CD48d (α4 integrin, which forms a heterodimer with CD29 to create very late activation antigen-4, VLA-4), while cells derived from the bone marrow do not. The pattern of expression of CD106 (VCAM-1), a VLA-4 ligand, is reversed. Interestingly, it has been observed an inverse pattern of expression of VLA-4 and its cognate receptor VCAM-1 in adipose tissue and bone marrow-derived cells. Thus, ADSCs express CD106/VCAM-1 but not CD49d/VLA-4, while BMSCs express CD49d/VLA-4 but not VCAM-1 (Morizonoc et al. 2003). Similarly, ADSCs express high levels of CD54 (ICAM-1), while BMSCs have minimal expression of this molecule. This is an intriguing observation, given the important role of the interaction between these molecules in stem cell homing to and mobilization from the bone marrow (Papayannopoulou et al. 2001; Sudhoff and Sohngen 2002). By contrast, CD44, a hyaluronic acid/fibronectin receptor also implicated in hematopoietic stem cell adhesion, proliferation and mobilization, is expressed on both BMSCs and ADSCs (Kronenwett et al. 2000).

Yet, while the general the pattern of mesenchymal lineage differentiation in basal control medium of ADSCs and BMSCs is very similar, some differences have been observed upon incubation in differentiating medium. In osteogenic differentiation medium, alkaline phosphatase activity is significantly greater in ADSCs, while mineralization is more extensive within BMSCs. Both cell populations express mRNA specific for collagen type I, osteocalcin, osteonectin, osteopontin, bone morphogenetic protein(BMP)-1, parathyroid hormone receptor, retinoic acid X receptor(RXR) α, vitamin D and CBFA-1, a transcriptional factor that regulates multiple osteogenic genes (Zuk et al. 2001). The in vitro chondrogenesis from both ADSCs and BMSCs is very similar, with micromasses of cells from both deposits, producing a proteoglycan rich, collagen type II-containing extracellular matrix with chondroitin-4-sulfate and keratin sulphate, two predominant glycosaminoglycans expressed in cartilage proteoglycan (Zuk et al. 2001; Huang et al. 2002). Despite the differences listed above, the similarities between stem cells extracted from the bone marrow and the adipose tissue suggest the potential for adipose tissue to act as an alternate, perhaps preferred, cell source for clinical application (Fraser 2002). Therapeutic enhancement of neovascularization is one of the most important strategies needed to limit the complications of postischemic injury. Considerable efforts have focused on the development of therapeutic strategies designed to increase vessel growth in the setting of ischemia. Particularly, transplantation of bone marrow-derived mononuclear cells (BM-MNC) has been shown to stimulate neovascularization after experimental ischemic injury, resulting in long-term salvage and survival of viable tissue (Mendez-Ferrer et al. 2006). The use of BM-MNC is now under intense investigation in humans, and the results of early small and uncontrolled studies point to a great potential for such therapy to limit disease progression (Chien 2006; Bartunek et al. 2007). Several researchers have showed that the administration of adipose lineage cells can potentially affect revascularization to a similar extent as BM-MNC administration (Miranville et al. 2004; Planat-Benard et al. 2004; Nakagami et al. 2006). Nakagami and collegues pointed out on the capability of ADSCs to secrete multiple angiogenic growth factors, such as VEGF and hepatocyte growth factor (HGF) (Nakagami et al. 2005). They showed that ADSCs secrete angiogenic growth factors that strongly induce endothelial growth, migration and tube formation. Transplantation of these cultured ADSCs to ischemic limbs accelerated angiogenesis mainly caused by secretion of growth factors, rather that participation in vessel formation by differentiation. Recent reports demonstrated that these ADSCs can be induced to differentiate into cardiac myocytes, which could be excellent for regenerative cell therapy for severe heart failure (Rangappa et al. 2003; Planat-Benard et al. 2004). The first and consistent report so far on the in vivo use of ADSCs in cardiac cell transplantation is from Miyahara et al. (2006). Miyahara and collegues (2006), using cell sheet technology, demonstrated that monolayered adipose tissue derived mesenchymal stem cells have multipotent and self-propagating properties after transplantation into infarcted rat hearts. Four weeks after coronary ligation, they transplanted the monolayered mesenchymal stem cells (MSCs) onto the scarred myocardium. After transplantation, the engrafted sheet gradually grew to form a thick stratum that included newly formed vessels, undifferentiated cells and few cardiomyocytes. The mesenchymal stem cell sheet also acted through paracrine pathways to trigger angiogenesis. They have also shown that transplantation of the monolayered MSCs significantly increased left ventricule maximum diastolic velocity (–dp/dt max), decreased the left ventricule end-diastolic pressure (LVEDP) and inhibited the development of left ventricle enlargement in rats with chronic heart failure secondary to myocardial infarction. These results suggest that transplantation of adipose tissue-derived monolayered MSCs improves cardiac function. However in this study the presence of cardiomyocytes within the MSC tissue seemed to be rare. Thus, the improvement of left ventricule function described here may be explained mainly by growth factor-mediated paracrine effects of the MSC sheet and a decrease in left ventricle wall stress resulting from the thick MSC. Although all these numerous studies have provided evidence that adipose tissue-derived stromal cells (ADSCs) contain a population of adult multipotent mesenchymal stem cells capable to offer a therapeutical potential for repairing damaged tissues, such as the ischemic myocardium, a number of fundamental issues need to be addressed before this approach can be considered for translational studies. First, although the majority of patients with acute myocardial infarction (MI) undergo spontaneous or iatrogenic reperfusion, the ability of ADSCs to improve cardiac function in the setting of a reperfused infarction has never been tested. The literature is replete with examples of therapies that work in the presence of a transient but not permanent coronary occlusion. Consequently, it is of utmost importance to determine whether ADSCs are effective when coronary occlusion is followed by reperfusion, an event that dramatically alters the milieu of the myocardial interstitium and of the myocardium itself. Second, improvement in left ventricular function has been shown when stem cells are injected intramyocardially in the peri-infarct area, an approach that would be difficult in patients. Clinically, the most practical route for ADSC administration is intravascular delivery, but it is unknown whether ADSCs injected into the coronary circulation can cross the vessel wall, translocate to the infarcted region, and initiate effective myocardial regeneration. Third, the mechanism whereby ADSCs improve cardiac function (differentiation into cardiac cells vs. fusion vs. paracrine effects on preexisting cells) remains poorly understood. Finally, the advantage of using total unfractionated ADSCs—compared with purified populations of ADSCs, such as the $CD34^+$-purified ADSCs, remains to be determined.

Major limitations in the use of adult stem cells harvested from essential organs are the pain, expense associated with the procedures, and the risk of donor site tissue morbidity. For the bone marrow, traditional tissue procurement procedure frequently requires general or spinal anesthesia and may yield low numbers of stem cells upon processing (approximately 1 bone marrow-derived stem cell per 10^5 adherent stromal cells) (Rickard et al. 1996; Pittenger et al. 1999). Studies employing human bone marrow cell transplantation in ischemic patients suggest that human angiogenic cell therapy requires at least 10^7–10^9 cells, depending on the degree of stem cells purity as well as the optimal delivery method. For the adult blood or umbilical cord, the use of endothelial progenitor cells (EPCs) for neovasculogenesis-based therapy would require large number of EPCs, but the rarity of these progenitors in peripheral blood would require large quantities of patient blood to isolate sufficient cells to achieve a pro-angiogenic or a pro-neovasculogenic effect (Orlic et al. 2001). Such quantities of blood are not readily available in a clinical setting. A comparative analysis of stem cells obtained from bone marrow, adipose tissue, and umbilical cord clearly showed that ADSCs were not different regarding morphology, immune phenotype, success rate of stem cells isolation, colony frequency, and differentiation capacity (Izadpanah et al. 2006; Kern et al. 2006). In a comparative study of BMSCs and ADSCs obtained from the same donors, De Ugarte et al. (Morizonoc et al. 2003) demonstrated that ADSCs required approximately 5 % the cell number used for marrow cells in order to reach initial confluence by 1 week. This suggests an increased proliferative potential, either by virtue of properties intrinsic to the cells or as the result of a greater frequency of stem cells within population used to initiate the culture. Thus, assuming clinical equivalence, the higher proliferative activity of the adipose tissue-derived population will generate a clinically effective cell dose more rapidly than the same number of marrow cells. Recent studies from Puissant and collegues (2005) reported on the in vivo and in vitro immunosuppressive properties of ADSCs. ADSCs have also been shown to escape the immune recognition because they do not express major histocompatibility complex (MHC) class II antigens, and, consequently, they do not induce allospecific T cell proliferative responses. These properties strengthen the clinical relevance of ADSCs in allogenic transplantation by reducing the incidence and severity of graft-versus-host disease. Yet, ADSCs from allogenic healthy donor could constitute a valuable alternative source of stem cells for therapeutic use in older patients or those with malignant diseases, who cannot yield sufficient numbers of functional ADSCs.

4.4 Cardiac Stem Cells

c-kit. Among stem cell populations, one of the best characterized was first isolated by Beltrami et al. by virtue of its expression of a c-Kit (stem cell factor receptor) surface marker (Beltrami et al. 2003). This population was found to be negative for blood lineage markers and heterogeneously positive for typical markers of cardiomyocyte commitment. They may undergo further expansion and, furthermore, demonstrate self-renewal and clonogenic capacity. In addition, c-Kit positive cells are multipotent, since capable to differentiate into cardiomyocytes, smooth muscle cells and endothelial cells (Barile et al. 2007). Nonetheless, these cells demonstrate a remarkable regenerative potential when injected into rat hearts following surgically induced myocardial infarction (MI) (Beltrami et al. 2003). In this particular study, injecting labelled cells facilitated the identification of myocardial regeneration in 19/20 treated hearts. Although it was reported that these cardiomyocytes were smaller, they also expressed sarcomeric protein, connexin 43 and, most notably, they exhibited visible striations. In addition, in vivo differentiation also generated both endothelial and smooth muscle cells, as evidenced by labelling of blood vessels cells. Importantly, subsequent investigations confirmed that such labeled cardiomyocytes were not the result of cell fusion (Beltrami et al. 2003). In the human heart, the atria are richest in c-Kit+ cardiac progenitors (Castaldo et al. 2008), followed by the left ventricle, right ventricle, apex and atrioventricular junction (Fuentes and Kearns-Jonker 2013). Disease and gender influence the distribution of these cells (Chan et al. 2012; Fuentes and Kearns-Jonker 2013). For example, females have a greater number of c-Kit+ cardiac progenitors (Itzhaki-Alfia et al. 2009). Moreover, an individual's medical history, including the use of β-blockers, smoking, atrial fibrillation and previous MI, can influence the abundance of c-Kit+ progenitors in the heart (Gambini et al. 2012).

Cardiosphere-derived cardiac stem cells. Resident cardiac stem cell populations can be isolated from biopsies of human and murine heart in primary culture (Messina et al. 2004). The method for isolation and clinical use of these cells is as follows: an endomyocardial biopsy (usually from the left atrial appendage or a ventricle) is first partially digested enzymatically. The subsequent sample is plated on fibronectin-coated dishes and kept in culture until it forms a layer of cells that have migrated out of the tissue (Smith et al. 2007). Human-derived cells must be cultured for 2–7 days, while it takes 14–24 days for porcine cells (Barile et al. 2007). Once confluent, these cells are harvested and plated on poly-D-lysine coated dishes in the presence of cardiosphere-forming medium. Thereafter, the cells develop in sphere-like structures, termed primary cardiospheres (Smith et al. 2007). When subsequently trypsinized and passaged as a monolayer, the cells form cardiosphere-derived cells (CDCs). CDCs that are exposed to epidermal growth factor (EGF) can, in turn, form further cardiospheres, which are termed secondary cardiospheres (Chimenti et al. 2010). At the centre of cardiosphere there is a core of primitive stem cells expressing c-Kit, nkx2.5 and desmin (Fuentes and Kearns-Jonker 2013), while the outer layers are composed of cells that are increasingly differentiated towards both cardiomyocyte and endothelial cells, as well as stromal mesenchymal cells (Fuentes and Kearns-Jonker 2013). Immunohistochemical analysis shows that the outer layers of cardiospheres express connexin 43 (the main gap junction protein in the heart (Urbanek et al. 2006)), sarcomeric proteins including troponin I and myosin heavy chain, and vascular proteins such as kinase domain receptor (Smith et al. 2008). The presence of gap junctions between uncommitted and differentiated cells within cardiospheres suggests that the differentiated cells may have a supporting role (Urbanek et al. 2006). The c-Kit+ population increases as cellular proliferation occurs within the core of the sphere, as evidenced by BrdU incorporation (Smith et al. 2008). Primary cardiospheres secrete a wide array of factors, including vascular endothelial growth factor (VEGF), hepatocyte growth factor and insulin-like growth factor 1, while CDCs and secondary cardiospheres secrete only VEGF (Chimenti et al. 2010). As with other types of stem cell, following characterization, in vivo experiments were performed to evaluate their regenerative capacity after an ischemic event. In a mouse model of MI both syngeneic and allogeneic transplanted CDCs were, without immunosuppression, able to significantly improve infarct size, infarcted wall thickness and ejection fraction compared to controls (Malliaras et al. 2012).

References

Assmus B, Schachinger V et al (2002) Transplantation of progenitor cells and regeneration enhancement in acute myocardial infarction (TOPCARE-AMI). Circulation 106(24):3009–3017

Bacou F, el Andalousi RB et al (2004) Transplantation of adipose tissue-derived stromal cells increases mass and functional capacity of damaged skeletal muscle. Cell Transplant 13(2):103–111

Barile L, Messina E et al (2007) Endogenous cardiac stem cells. Prog Cardiovasc Dis 50(1):31–48

Barry F, Boynton R et al (2001) The SH-3 and SH-4 antibodies recognize distinct epitopes on CD73 from human mesenchymal stem cells. Biochem Biophys Res Commun 289(2):519–524

Bartunek J, Vanderheyden M et al (2007) Bone-marrow-derived cells for cardiac stem cell therapy: safe or still under scrutiny? Nat Clin Pract Cardiovasc Med 4(Suppl 1):S100–S105

Beltrami AP, Barlucchi L et al (2003) Adult cardiac stem cells are multipotent and support myocardial regeneration. Cell 114(6):763–776

Britten MB, Abolmaali ND et al (2003) Infarct remodeling after intracoronary progenitor cell treatment in patients with acute myocardial infarction (TOPCARE-AMI): mechanistic insights from serial contrast-enhanced magnetic resonance imaging. Circulation 108(18):2212–2218

Castaldo C, Di Meglio F et al (2008) CD117-positive cells in adult human heart are localized in the subepicardium, and their activation is associated with laminin-1 and alpha6 integrin expression. Stem Cells 26(7):1723–1731

Chan HH, Meher Homji Z et al (2012) Human cardiosphere-derived cells from patients with chronic ischaemic heart disease can be routinely expanded from atrial but not epicardial ventricular biopsies. J Cardiovasc Transl Res 5(5):678–687

Chien KR (2006) Lost and found: cardiac stem cell therapy revisited. J Clin Invest 116(7): 1838–1840

Chimenti I, Smith RR et al (2010) Relative roles of direct regeneration versus paracrine effects of human cardiosphere-derived cells transplanted into infarcted mice. Circ Res 106(5):971–980

Fraser JK (2002) Adipose tissue: challenging the marrow monopoly. Cytotherapy 4(6):509–510

Fuchs S, Satler LF et al (2003) Catheter-based autologous bone marrow myocardial injection in no-option patients with advanced coronary artery disease: a feasibility study. J Am Coll Cardiol 41(10):1721–1724

Fuentes T, Kearns-Jonker M (2013) Endogenous cardiac stem cells for the treatment of heart failure. Stem Cells Cloning 6:1–12

Gambini E, Pesce M et al (2012) Patient profile modulates cardiac c-kit(+) progenitor cell availability and amplification potential. Transl Res 160(5):363–373

Gronthos S, Graves SE et al (1994) The STRO-1+ fraction of adult human bone marrow contains the osteogenic precursors. Blood 84(12):4164–4173

Hicok KC, Laney TVD et al (2004) Human adipose-derived adult stem cells produce osteoid in vivo. Tissue Eng 10(3–4):371–380

Huang JI, Beanes SR et al (2002) Rat extramedullary adipose tissue as a source of osteochondrogenic progenitor cells. Plast Reconstr Surg 109(3):1033–1041; discussion 1042–1043

Itzhaki-Alfia A, Leor J et al (2009) Patient characteristics and cell source determine the number of isolated human cardiac progenitor cells. Circulation 120(25):2559–2566

Izadpanah R, Trygg C et al (2006) Biologic properties of mesenchymal stem cells derived from bone marrow and adipose tissue. J Cell Biochem 99(5):1285–1297

Kalka C, Masuda H et al (2000) Transplantation of ex vivo expanded endothelial progenitor cells for therapeutic neovascularization. Proc Natl Acad Sci U S A 97(7):3422–3427

Kamihata H, Matsubara H et al (2001) Implantation of bone marrow mononuclear cells into ischemic myocardium enhances collateral perfusion and regional function via side supply of angioblasts, angiogenic ligands, and cytokines. Circulation 104(9):1046–1052

Kern S, Eichler H et al (2006) Comparative analysis of mesenchymal stem cells from bone marrow, umbilical cord blood, or adipose tissue. Stem Cells 24(5):1294–1301

Kronenwett R, Martin S et al (2000) The role of cytokines and adhesion molecules for mobilization of peripheral blood stem cells. Stem Cells 18(5):320–330
Malliaras K, Li TS et al (2012) Safety and efficacy of allogeneic cell therapy in infarcted rats transplanted with mismatched cardiosphere-derived cells. Circulation 125(1):100–112
Mendez-Ferrer S, Ellison GM et al (2006) Resident progenitors and bone marrow stem cells in myocardial renewal and repair. Nat Clin Pract Cardiovasc Med 3(Suppl 1):S83–S89
Messina E, De Angelis L et al (2004) Isolation and expansion of adult cardiac stem cells from human and murine heart. Circ Res 95(9):911–921
Miranville A, Heeschen C et al (2004) Improvement of postnatal neovascularization by human adipose tissue-derived stem cells. Circulation 110(3):349–355
Miyahara Y, Nagaya N et al (2006) Monolayered mesenchymal stem cells repair scarred myocardium after myocardial infarction. Nat Med 12(4):459–465
Morizonoc K et al (2003) Comparison of multi-lineage cells from human adipose tissue and bone marrow. Cells Tissues Organs 174:101–109
Nakagami H, Maeda K et al (2005) Novel autologous cell therapy in ischemic limb disease through growth factor secretion by cultured adipose tissue-derived stromal cells. Arterioscler Thromb Vasc Biol 25(12):2542–2547
Nakagami H, Morishita R et al (2006) Adipose tissue-derived stromal cells as a novel option for regenerative cell therapy. J Atheroscler Thromb 13(2):77–81
Orlic D, Kajstura J et al (2001) Bone marrow cells regenerate infarcted myocardium. Nature 410(6829):701–705
Papayannopoulou T, Priestley GV et al (2001) Molecular pathways in bone marrow homing: dominant role of alpha(4)beta(1) over beta(2)-integrins and selectins. Blood 98(8):2403–2411
Perin EC, Silva GV et al (2002) Assessing myocardial viability and infarct transmurality with left ventricular electromechanical mapping in patients with stable coronary artery disease: validation by delayed-enhancement magnetic resonance imaging. Circulation 106(8):957–961
Perin EC, Dohmann HF et al (2003) Transendocardial, autologous bone marrow cell transplantation for severe, chronic ischemic heart failure. Circulation 107(18):2294–2302
Pittenger MF, Mackay AM et al (1999) Multilineage potential of adult human mesenchymal stem cells. Science 284(5411):143–147
Planat-Benard V, Menard C et al (2004) Spontaneous cardiomyocyte differentiation from adipose tissue stroma cells. Circ Res 94(2):223–229
Puissant B, Barreau C et al (2005) Immunomodulatory effect of human adipose tissue-derived adult stem cells: comparison with bone marrow mesenchymal stem cells. Br J Haematol 129(1):118–129
Rangappa S, Fen C et al (2003) Transformation of adult mesenchymal stem cells isolated from the fatty tissue into cardiomyocytes. Ann Thorac Surg 75(3):775–779
Rickard DJ, Kassem M et al (1996) Isolation and characterization of osteoblast precursor cells from human bone marrow. J Bone Miner Res 11(3):312–324
Simmons PJ, Torok-Storb B (1991) Identification of stromal cell precursors in human bone marrow by a novel monoclonal antibody, STRO-1. Blood 78(1):55–62
Smith RR, Barile L et al (2007) Regenerative potential of cardiosphere-derived cells expanded from percutaneous endomyocardial biopsy specimens. Circulation 115(7):896–908
Smith RR, Barile L et al (2008) Stem cells in the heart: what's the buzz all about? Part 2: arrhythmic risks and clinical studies. Heart Rhythm 5(6):880–887
Strauer BE, Brehm M et al (2002) Repair of infarcted myocardium by autologous intracoronary mononuclear bone marrow cell transplantation in humans. Circulation 106(15):1913–1918
Sudhoff T, Sohngen D (2002) Circulating endothelial adhesion molecules (sE-selectin, sVCAM-1 and sICAM-1) during rHuG-CSF-stimulated stem cell mobilization. J Hematother Stem Cell Res 11(1):147–151
Urbanek K, Cesselli D et al (2006) Stem cell niches in the adult mouse heart. Proc Natl Acad Sci U S A 103(24):9226–9231
Zuk PA, Zhu M et al (2001) Multilineage cells from human adipose tissue: implications for cell-based therapies. Tissue Eng 7(2):211–228

Part II
Critical Regulators of Stem Cell Differentiation for Myocardial Regeneration

Chapter 5
Circadian Rhythms in Stem Cell Biology and Function

Pieterjan Dierickx, Bastiaan Du Pré, Dries A.M. Feyen, Niels Geijsen, Toon van Veen, Pieter A. Doevendans, and Linda W. Van Laake

5.1 Introduction

Terrestrial life revolves around the 24-h cycle of day and night. The light-dark cycle has a direct influence on organismal functioning, dictating wake-sleep patterns in animals and cycles of photosynthesis in plants. An underlying mechanism termed the circadian clock regulates these processes at the molecular level. The word circadian is derived from the two Latin words *circa* and *dies* which mean "around" and "day" respectively. This mechanism is heavily conserved throughout evolution and allows organisms to adapt and to synchronize themselves to diurnal fluctuations in their environment. Circadian rhythmicity can be seen in many different life forms, ranging from unicellular organisms, like cyanobacteria, to highly specialized and

P. Dierickx • L.W. Van Laake (✉)
Division of Heart and Lungs, Department of Cardiology, University Medical Center Utrecht, Utrecht, The Netherlands

Hubrecht Institute-KNAW and University Medical Center, Utrecht, The Netherlands
e-mail: L.W.vanLaake@umcutrecht.nl

B. Du Pré
Hubrecht Institute-KNAW and University Medical Center, Utrecht, The Netherlands

Division of Heart and Lungs, Department of Medical Physiology, University Medical Center Utrecht, Utrecht, The Netherlands

D.A.M. Feyen
Division Heart and Lung, Cardiology, Experimental Cardiology Laboratory, University Medical Center Utrecht, Utrecht, The Netherlands

N. Geijsen
Hubrecht Institute-KNAW and University Medical Center, Utrecht University, Utrecht, The Netherlands

Faculty of Veterinary Medicine, Clinical Sciences of Companion Animals, Utrecht University, Utrecht, The Netherlands

R. Madonna (ed.), *Stem Cells and Cardiac Regeneration*, Stem Cell Biology and Regenerative Medicine, DOI 10.1007/978-3-319-25427-2_5

complex multicellular organisms, such as mammals. Concomitantly, circadian rhythms regulate many body features of animals like behavior, metabolism, blood pressure, body temperature, tissue physiology, regeneration and homeostasis (Aschoff 1983).

The central core clock in mammals is located in the suprachiasmatic nucleus (SCN), a group of approximately 20,000 neurons in the anterior hypothalamus in the brain. The SCN clock is driven by light, the signal that is relayed after perception in the retina by photoreceptors. However, the clock processes are not driven by light per se. Light should be seen as the main external synchronizer (*also known as zeitgeber; time-giver*) forcing the body to adapt to a 24-h period, rather than driving circadian rhythmicity in physiology directly. The SCN synchronizes peripheral clocks through neural and humoral factors like the serotonin-derived hormone melatonin (Cajochen et al. 2003). Peripheral clocks are present in almost all tissues in the mammalian body, including liver, lung, kidney, skin, and the heart. These clocks maintain circadian tissue physiology via controlling tissue-specific gene expression (Brown and Azzi 2013). The molecular machinery behind this timekeeping system comprises multiple genes, termed *clock genes*, which products interact with each other ensuring stable and robust circadian rhythmicity.

The hallmark of circadian rhythms is that they keep on cycling with the same phase in the absence of an external input. This can be seen by the persistence of circadian rhythmicity when animals are kept in complete darkness. Another typical feature of the circadian clock is the fact that it is not altered by external perturbation or at mild variations of ambient temperatures, a process known as "temperature compensation" (Merrow et al. 2005). This is nicely illustrated by the fact that the clocks of warm-blooded animals are buffered against and maintained at different temperatures throughout the day.

The importance of maintaining a functional time-keeping system is shown by the fact that disruption of the clock has been associated with a vast array of malignancies, such as impairment of lipid homeostasis resulting in a fatty liver and obesity (Adamovich et al. 2014). A disturbed regulation of the clock has also been linked to the development of cardiovascular diseases, multiple sleep disorders, depression, inflammation, cancer, impairment of regenerative capacity and other metabolic disorders (Rudic et al. 2004; Kennaway et al. 2013; Lumaban and Nelson 2014) like diabetes (Marcheva et al. 2010; Milagro et al. 2012). Furthermore, recent research has shown that circadian timekeeping can also be linked to developmental and physiological processes that are not necessarily associated with a 24-h daily pattern. This can be observed in clocks regulating somatic

T. van Veen
Division of Heart and Lungs, Department of Medical Physiology, University Medical Center Utrecht, Utrecht, The Netherlands

P.A. Doevendans
Division of Heart and Lungs, Department of Cardiology, University Medical Center Utrecht, Utrecht, The Netherlands

stem cell heterogeneity (Janich et al. 2011), cell division (Matsuo et al. 2003; Kowalska et al. 2013; Feillet et al. 2014; Nagoshi et al. 2004; Yang et al. 2009; Unsal-Kaçmaz et al. 2005), damage induced regeneration (Janich et al. 2013), immune progenitor cell migration and differentiation (Scheiermann et al. 2013; Yu et al. 2013) as reviewed by Steven Brown (2014).

In the heart specifically, the circadian clock modulates the response to induced damage, such as ischemia/reperfusion, as demonstrated by the fact that diurnal oscillations in infarct size are blunted in cardiomyocyte specific clock mutants (Durgan et al. 2010). However, the precise role of circadian clocks in cardiac regenerative medicine still needs to be determined. Stem cells can now efficiently be differentiated towards cardiac cells in vitro as reviewed in Dierickx et al. (2012). Studying how circadian rhythms in cultured cells can enhance their regenerative effects, but also elucidating the optimal administration time of these cells into the myocardium, will be areas of research that could boost the field towards more effective therapy.

5.2 The Core Clock Machinery

5.2.1 A Complex Transcriptional Feedback Loop Defines the Molecular Clock

The molecular clock machinery comprises an interlocking activating and inhibiting transcriptional/(post)translational feedback loop (TTFL). This renders a 24-h rhythmicity pattern in expression of clock-controlled genes (CCGs), resulting in a circadian functional output (schematic overview in Fig. 5.1). Two of the main players in this complex system are BMAL1 (Brain and Muscle ARNT-Like 1) and CLOCK (Circadian Locomotor Output Cycles Kaput), encoded by *Arntl* and *Clock* respectively. They are both bHLH-PAS (basic helix-loop-helix, *Per-Arnt-Single-minded*) proteins and form the center of the activating limb of the circadian clock pathway. Upon heterodimerization via their PAS domains, they drive gene expression via docking on two types of enhancer box elements (E-Boxes): E-box (5′-CACGTG-3′) and E′-box (5′-CACGTT-3′). These E-boxes lay near or in the promoter of their targets, termed CCGs (Hogenesch et al. 1997; Gekakis et al. 1998; Yoo et al. 2005; Ohno et al. 2007). It is now clear that both genes are inevitable to sustain a proper clock. For example, knocking out *Bmal1* results in the complete loss of behavioral rhythmicity (Bunger et al. 2000; Shi et al. 2010). *Clock* knockout mice do not show this phenotype, but this is probably because the role of CLOCK can be bypassed by NPAS2, a protein with an analogous function. Indeed, NPAS2 deficient mice still show rhythmic behavior (Dudley et al. 2003), which is lost in *Clock/Npas2* double knockouts (DeBruyne et al. 2007).

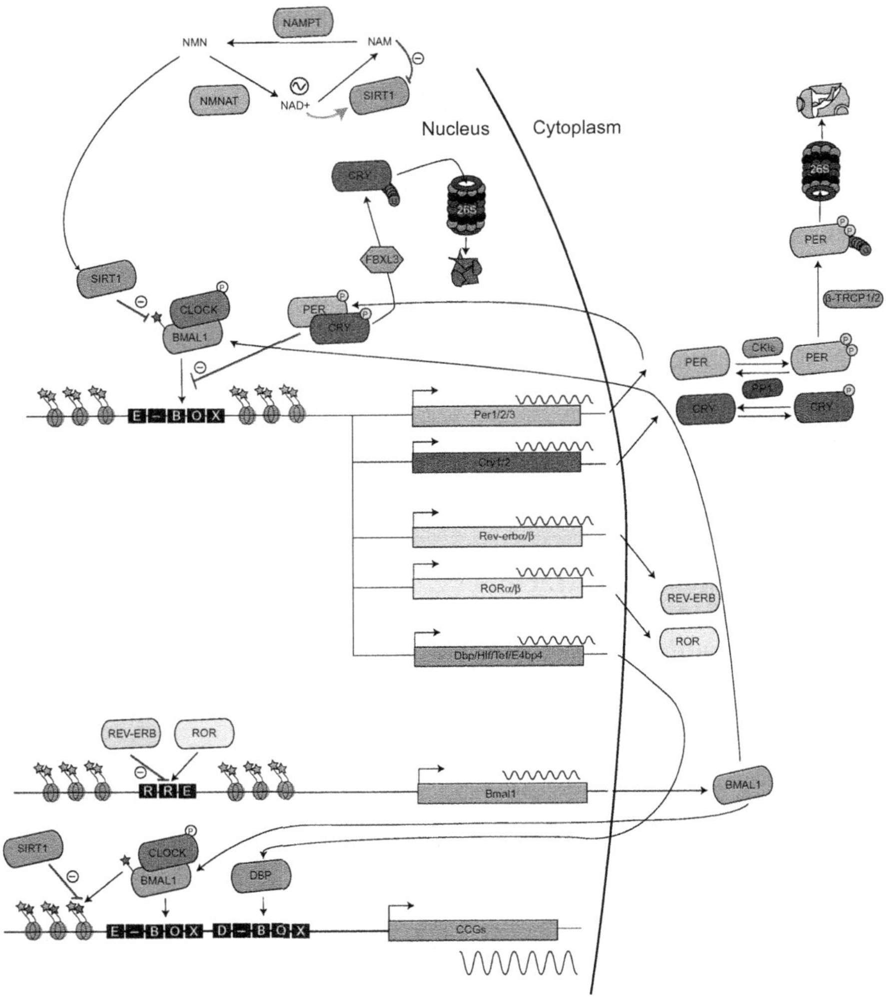

Fig. 5.1 Schematic representation of the transcriptional/translational feedback loop of the circadian core clock pathway in mammals. The central proteins BMAL1 and CLOCK form a heterodimer and bind E-box elements in the promoter of period (*Per1/2*) and cryptochrome (*Cry1/Cry2*) genes. PER and CRY can be degraded by the 26S proteasome in the cytoplasm after ubiquitination (U) by β-TRCP1/2 after casein kinase ε mediated phosphorylation (P). If not degraded in the cytoplasm, PER and CRY can dimerize and shuttle to the nucleus, where they inhibit their own transcription via blocking BMAL1:CLOCK transcriptional activity. After gradual phosphorylation, the PER/CRY complex is ubiquitinated by the F-Box protein FBXL3, and degraded in the nucleus. This lifts the repression on BMAL1:CLOCK, resulting in a new transcriptional cycle. In a secondary feedback loop, the BMAL1:CLOCK dimer drives transcription of *Rev-erbα/β* and *RORα/β*. Their proteins shuttle to the nucleus where they inhibit and activate *Bmal1* transcription, respectively via competing for a Rev-erb response element (RRE) in the *Bmal1* promoter. The result of one cycle, which takes approximately 24 h is the rhythmic transcriptional activation of clock-controlled genes (CCGs) by BMAL1:CLOCK. An oscillatory SIRT1 activity cycle integrates metabolism into the circadian clock. For additional information see text. *Stars*: histone tails

Transcription of two groups of CCGs, named Period (*Per*) genes and Cryptochrome (*Cry*) genes, are activated by BMAL1 and CLOCK. These output genes constitute the negative branch of the first autoregulatory feedback loop. Their necessary role in the clock pathway is underscored by the fact that *Cry1/Cry2* double knockout mice also show a complete loss of rhythmicity (van Der Horst et al. 1999; Vitaterna et al. 1999). PER and CRY proteins accumulate and dimerize in the cytoplasm, where their presence is tightly regulated. They can either be stabilized by phosphatase1 (PPI1) (Gallego et al. 2006) or phosphorylated by casein kinases (CKε/δ) (Keesler et al. 2000; Lowrey et al. 2000; Camacho et al. 2001) resulting in active degradation. For CRY1 this process occurs via ubiquitination by F-box and leucine-rich repeat protein 21 (FBXL21) (Dardente et al. 2008; Hirano et al. 2013; Yoo et al. 2013). Phosphorylated PER2 can be polyubiquitinated by β-TRCP1/2 resulting in proteasomal degradation (Eide et al. 2005). In general, dimerization of PER:CRY in the cytoplasm protects both proteins from degradation (Yagita et al. 2002). Upon stabilization, the PER:CRY dimer translocates into the nucleus (Yagita et al. 2000) forming a nuclear complex (Brown et al. 2005). There, the dimer binds to the NuRD (nucleosome remodeling deacetylases) transcriptional repressor complex and directs NuRD to BMAL1:CLOCK (Kim et al. 2014). Then, a fully functional NuRD repressing complex is established, resulting in the inhibition of BMAL1:CLOCK driven transcription. Through this, PER and CRY inhibit multiple CCGs as well as their own transcription (Gekakis et al. 1998; Kume et al. 1999; Griffin et al. 1999). PER initiates this negative feedback loop by functioning as a molecular scaffold, that brings CRY into contact with BMAL1:CLOCK (Chen et al. 2009). In the nucleus CRY1 and CRY2 can be degraded by the proteasome, facilitated by F-box type E3 ubiquitin ligase (FBXL3). Its function is nicely illustrated by impaired degradation of CRY, as seen in the overtime (Ovtm) mutant, that causes enhanced inhibition of BMAL1:CLOCK-based transcription, resulting in period lengthening (Siepka et al. 2007). Upon decrease of nuclear PER and CRY levels, BMAL1:CLOCK inhibition is released, and a new transcriptional activation cycle of CCGs can start.

An additional layer of transcriptional control of the circadian clock comprises the orphan nuclear receptors ROR α/β and REV-ERB α/β, encoded by *Ror* α/β and *Nr1d1/2* respectively. The BMAL1:CLOCK dimer mediates their transcription via binding to their E-box, initiating a second feedback loop (Preitner et al. 2002; Sato et al. 2004; Guillaumond et al. 2005). Both RORs and REV-ERBs compete for retinoic acid-related orphan receptor response element (RORE) binding sites, also termed RRE (RevErbA response element), within the *Bmal1* promoter (Harding and Lazar 1993; Ueda et al. 2002). Binding of RORs drives transcription of *Bmal1* while REV-ERBα inhibits its transcription (Preitner et al. 2002; Akashi and Takumi 2005; Guillaumond et al. 2005). Although *Rora* mRNA levels are only slightly oscillating, RORα is necessary for rhythmic *Bmal1* expression (Akashi and Takumi 2005).

A last transcriptional loop involved in circadian clock oscillation consists of a number of proline and acidic amino acid-rich basic leucine zipper (PAR bZIP) transcription factors regulating circadian gene expression via binding D-box (DBP response element) elements (TTATG(C/T)AA) (Falvey et al. 1996). Albumin D-site-binding protein (DBP), thyrotroph embryonic factor (TEF), and hepatic leukemia factor (HLF) contribute to positive regulation, whereas nuclear factor interleukin 3 regulated (NFIL3 or E4BP4) provides negative regulation. Although this accessory loop is not strictly necessary for circadian oscillations, it provides robustness and precision to the period. An overview of the described core clock pathway is depicted in Fig. 5.1.

5.2.2 Epigenetic Regulation of the Circadian Clock

Besides the complex transcriptional/translational control system, circadian rhythmicity is also regulated by epigenetic mechanisms. Epigenetic control includes methylation of the DNA at CpG islands, non-coding RNAs and posttranslational modifications of histones. All of these epigenetic mechanisms have been implicated in driving and fine-tuning circadian rhythmicity in gene expression.

CLOCK, one of the core circadian rhythm proteins, can function as a histone acetyl transferase (HAT) that acetylates histone H3 on its lysine 9 (H3K9) and lysine 14 (H3K14) amino acid residues (Doi et al. 2006). H3K9ac and H3K14ac are both markers for permissive transcription. Therefore, BMAL1:CLOCK also regulates transcription of CCGs via modifying their histones (Etchegaray et al. 2003). This function is being neutralized by several histone deacetylases (HDACs), as described in more detail by Steven Brown (2011).

In search for these balancing HDACs, Sirtuin 1 (SIRT1) was discovered by the group of Sassone-Corsi as a protein that counteracts the HAT function of CLOCK. In general, *Sirt1* is well studied in the context of aging, resistance to cellular stress, metabolism, inflammation and proliferation (Bordone and Guarente 2005). Additionally, SIRT1 is now known to deacetylate the proteins BMAL1, PER2 (Asher et al. 2008) and histone 3 (H3) on the promoter of clock output genes like *Dbp* (Nakahata et al. 2008). The deacetylating activity of SIRT1 is NAD+ (nicotinamide adenine dinucleotide) dependent and circadian. In the absence of de novo NAD+ biosynthesis, NAD+ needs to be replenished to avoid cell death. This goes via the NAD+ salvage pathway, where the by-product of NAD+ usage, NAM (nicotinamide) is reconverted into usable NAD+ via NMN (nicotinamide mononucleotide). In this process, NAMPT (NAM phosphoribosyltransferase) is the rate-limiting enzyme. As the expression of NAMPT itself is under circadian control, NAD+ also oscillates as an available metabolite. By this, SIRT1 links the metabolic state of a cell with the epigenetic control of the clock gene transcription pathway. As a rheostat of the circadian clock, SIRT1 mainly controls the amplitude of CCGs (Sassone-Corsi 2012).

Besides acetylation, methylation of histones is important to mediate circadian rhythmicity in gene expression. Histone H3 lysine trimethylation (H3K4me3) is a mark consistently associated with circadian transitions of the chromatin fiber, controlling CCG expression. One of these histone methyl transferases (HMTs) is mixed lineage leukemia 1 (MLL1) that can recruit the BMAL1:CLOCK dimer to the DNA of target genes and cause rhythmic H3K4 trimethylation (Katada and Sassone-Corsi 2010). Histone H3 lysine 27 trimethylation (H3K27me3) on the other hand is a repressive epigenetic mark and has been shown to play a counteracting role in circadian regulation of gene expression. The *Per1* promoter for example shows rhythmic H3K27me3 marks, mediated by EZH2. The counteracting enzymes, histone demethylases, also play a role in circadian rhythms, where JARID1A is a known demethylase that inhibits HDAC1 and boosts BMAL1:CLOCK facilitated transcription of *Per* genes. Dynamic interaction between HDAC1 and JARID1A correlates with proper histone acetylation at the *Per* promoters (DiTacchio et al. 2011). The same holds true for LSD1, another histone demethylase whose activity depends on circadian phosphorylation by PKCα (Nam et al. 2014).

5.2.3 *Additional Regulatory Systems Fine-Tune Circadian Rhythmicity*

As described above, the genetic and epigenetic mechanisms underlying the circadian clock are quite complex and have been studied extensively. In addition to all this, a handful of papers describe even more ways of circadian regulation of the core clock pathway. Most of these additional mechanisms enhance robustness of the clock, rather than being truly essential for rhythmicity. A first example of such an extra clock dimension is the oscillating cold inducible RNA-binding protein (CIRP). CIRP is a RNA binding protein, and is regulated by circadian rhythmicity in temperature oscillations. CIRP binds to Clock mRNA and stabilizes it, linking temperature to the circadian clock (Morf et al. 2012). Next, cyclic alternative splicing (McGlincy et al. 2012) as well as light inducible alternative splicing (Preußner et al. 2014) are common factors that regulate circadian rhythms. Third, rhythmic polyadenylation that stabilizes mRNA molecules, facilitates circadian rhythmicity in protein translation (Kojima et al. 2012). Furthermore, fluctuations in m(6)A-RNA methylation also affect the circadian transcriptome (Fustin et al. 2013). Finally, non-coding RNAs such as anti-sense RNAs and miRNAs affect circadian rhythms. This is demonstrated by a disrupted circadian transcriptome in cells that lack *Dicer*, a gene involved in miRNA processing (Chen et al. 2013; Du et al. 2014). In conclusion, the circadian clock is achieved by a complex orchestrated interplay between genetics, epigenetics and translational processes.

5.3 Circadian Rhythms in Tissue/Organ Physiology

5.3.1 *Tissue Specific Control of Clock Output*

The core clock machinery, as studied extensively in the SCN, is conserved in all peripheral organs. The basic signaling between the core oscillator and peripheral clocks involves both neuronal and humoral signals, such as melatonin. However, peripheral clocks can also respond independently to environmental cues such as body temperature and food metabolites (Damiola et al. 2000; Stokkan et al. 2001) (Fig. 5.2). Nonetheless, there are significant differences between each tissue in the

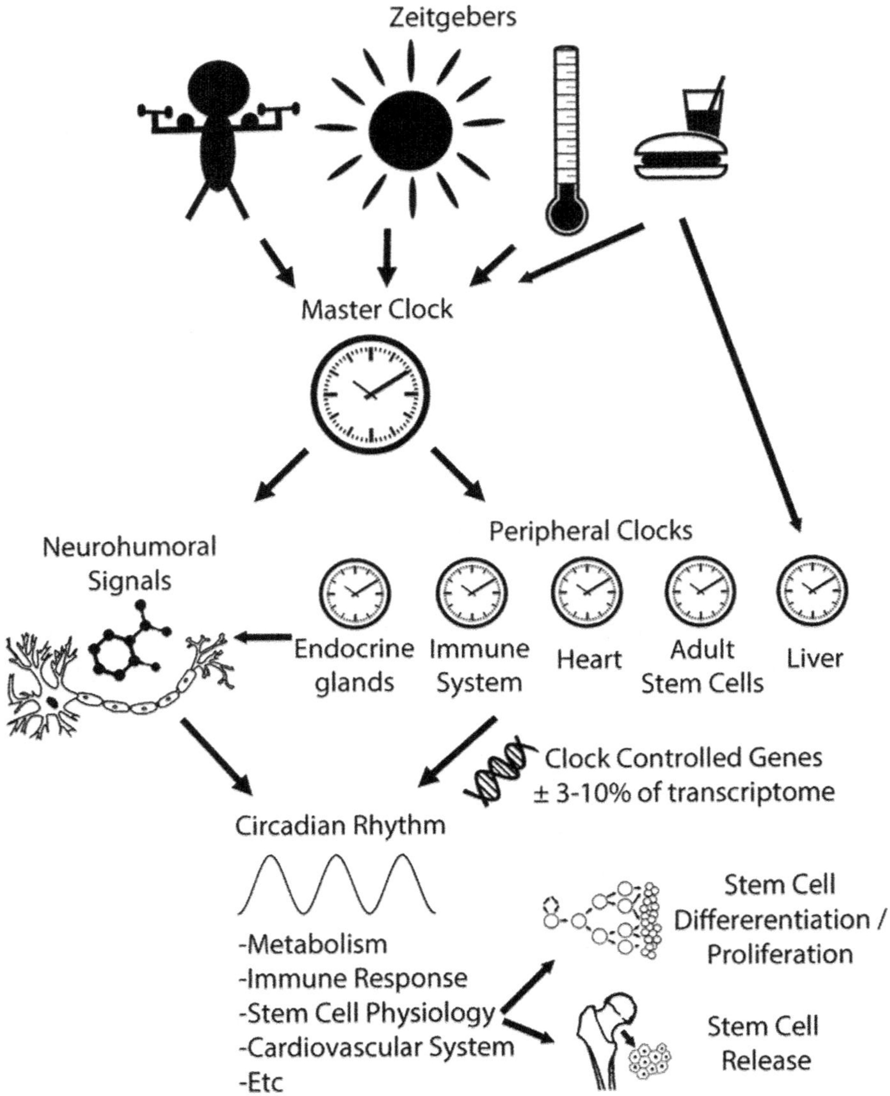

Fig. 5.2 The circadian clock regulates rhythmic body physiology. A master clock in the brain and peripheral clocks in almost every cell/tissue of the body drive circadian rhythmicity. Clocks are entrained by different factors/*zeitgebers* such as light and food. These rhythms are propagated by electrical and neurohumoral signals resulting in a 24-h rhythmic expression of clock-controlled genes. These genes render circadian rhythmicity in many functional processes ensuring proper body physiology and regeneration

relative contributions of the clock components, as well as in the output pathways that are under their control. These endogenous cellular clocks drive extensive rhythms of gene transcription, with 3–10 % of all mRNAs in a given tissue showing diurnal rhythms (Akhtar et al. 2002; Duffield et al. 2002; Miller et al. 2007). However, the genes that are under circadian control are largely non-overlapping in different tissues. This tissue specificity reflects the need for temporal control of the cellular physiology relevant to each unique cell type. As a result, the circadian clock exerts extensive control over many unique biological processes.

5.3.2 Transgenic Animals Provide Novel Insights to Better Understand the Clock

Tissue-specific transgenic mice have been used to address the precise functions of peripheral clocks in physiological processes. For example, in liver-specific *Bmal1* KO mice there is a loss of rhythmicity of glucose regulatory genes, which leads to an accelerated glucose clearance during the course of the daily feeding cycle (Lamia et al. 2008). In the adrenal glands, many genes involved in the biosynthesis of corticosterone are clock-controlled. Therefore the tissue specific disruption of *Bmal1* interrupts the ability of the organ to maintain proper oscillatory secretion of corticosterone (Son et al. 2008). In pancreatic islets, the circadian clock helps regulating glucose-stimulated insulin secretion, the loss of which impairs glucose tolerance because of β-cell dysfunction (Marcheva et al. 2010).

The peripheral clock also plays a profound role in the cardiovascular system (Durgan and Young 2010; Paschos and FitzGerald 2010). In blood vessels, regulation of vascular function and tone has shown to be regulated by circadian rhythms. For example, deletion of *Bmal1* specifically in vascular endothelium leads to a reduction of blood pressure during the active phase of the day and increased heart rate throughout the 24-h cycle (Westgate et al. 2008). The ability of endothelial cells from *Per2* mutant mice to proliferate and form vascular networks is substantially reduced, which is marked by increased senescence of the cells (Wang et al. 2008). In vivo, *Per2* mutant mice show decreased angiogenesis, as blood flow is impaired and combined with reduced recovery in response to ischemia characterized by a smaller increase in vessel formation. Lastly, *Westgate et al.* (2008) showed that the time to thrombotic vascular occlusion in response to a photochemical injury displays diurnal variation. Platelet aggregation factors, plasminogen activator inhibitor and tissue plasminogen activator, produced by the vascular endothelium show diurnal variability throughout the day/night cycle. In this regard, the deletion of *Bmal1* specifically in the vascular endothelium results in loss of the temporal pattern in susceptibility to thrombotic vascular occlusion.

5.3.3 *Circadian Rhythms in the Heart*

In the heart, rhythmic physiology has mostly been studied in mice harboring a dominant negative version of the CLOCK protein in cardiomyocytes specifically (CCM mice). In vivo radiotelemetry studies performed in wildtype (WT) and CCM mice for continuous 24-h monitoring of physical activity, revealed a reduction in heart rate in the CCM mice (despite identical physical activity), which was especially pronounced during the awake/active phase. In this regard, the circadian clock seems to influence the generation and propagation of electric signals between adjacent cells in the heart. Although the precise mechanism remains unknown for this phenotype, the expression of various ion channels was found to be clock dependent. Furthermore, Connexin 40, a gap junction protein critical in atrial-ventricular conduction, shows oscillatory expression in WT hearts, that is absent in CCM hearts (Bray et al. 2008).

The metabolism of the heart is crucial for its contractile function, which has to meet the daily demand for increased workload during the active phase of the day. The two major components fueling the contraction of the myocardium are fatty acids and glucose (Taegtmeyer 2000). Diurnal gene expression studies in CCM mice revealed that a large number of genes influencing triglyceride (fatty acids) and glycogen (glucose) metabolism are controlled by the clock (Bray et al. 2008). The circadian regulation of lipolysis is exhibited in the diurnal variation of total triglyceride levels and synthesis in mouse hearts that peaks near the end of the active phase. These rhythms are essentially absent in CCM hearts. Furthermore, CCM mice display an altered response to high fat diet, showing a role of the cardiomyocyte circadian clock in the regulation of nonoxidative fatty acid metabolism (Tsai et al. 2010). Similarly, epinephrine-induced glycogenolysis has a time of the day dependent activity in WT hearts. However, these diurnal rhythms in cardiac glycogen metabolism are suppressed in CCM hearts (Bray et al. 2008).

Furthermore, hearts from 22-month-old CCM mice exhibited increased ejection fraction, fractional shortening, and left ventricular mass compared to WT. These are all characteristic of physiological hypertrophy and a strong link that the intrinsic clock of the heart also regulates myocardial growth. Growth factors, such as insulin-like growth factor-1 (IGF-1), are at the basis of physiological hypertrophy and can signal through the PI3K-Akt pathway. In this regard, the phosphorylation status of key components of this signaling pathway (Akt, GSK3β, and p70S6K) all oscillate in hearts over the course of the day, and are continually elevated in CCM hearts (Durgan et al. 2010). Myocardial growth is also closely associated with protein synthesis, which involves initiation factors (eIFs), several of which have been found to be under cardiomyocyte circadian regulation (Bray et al. 2008).

In summary, the peripheral clocks allow for individualized rhythmic gene expression in order for organs to be able to anticipate their diurnal tasks. In the cardiac context, heart rate, metabolism, and growth are all parameters that are under tissue specific circadian control. This enables the heart to be used most efficiently during periods of activity and rest.

5.4 Ontology of Circadian Rhythms and the Role of Clocks in Cell Cycle

5.4.1 Circadian Clocks During Embryonic Development

While the embryo develops, conditions within the uterus vary throughout the day. Concentrations of glucose and other metabolites for example, are relatively high during daytime compared to at night. A mature circadian clock to anticipate to these diurnal differences is not yet present in the embryo, but develops during gestation.

The proteins of the circadian clock, including CLOCK, BMAL, PER, and CRY are already present in the unfertilized oocyte (Johnson et al. 2002). However, their expression is non-rhythmic and low compared to adult cells. During development *in utero*, the expression of those genes gradually increases until birth (Saxena et al. 2007). Around the period of mid- to end gestation, diurnal oscillations of the core clock proteins commence. The amplitude of expression increases until the end of pregnancy and upon delivery, there is a 12-h phase shift, reversing 24-h clock gene expression (Saxena et al. 2007). In some species, such as rodents, development of the molecular clock even continues after birth.

Circadian clocks rely on Zeitgebers for synchronization to external surroundings. Light for example, is received by the retina and via internal cues, synchronizes circadian clocks within the body. The fetus does not directly receive light input. Also, many of the maternal Zeitgeber signals do not reach the fetus because they do not pass the placenta. Therefore, the regulation of molecular clocks within the fetus rely mainly on melatonin, which is produced by the maternal pineal gland and is able to pass the placenta (Reppert and Weaver 2002). The obtained 24-h rhythms play an important role in the physiology of the developing fetus. Metabolic activity and parameters such as breathing movements and heart rate show diurnal oscillations within the uterus (Visser et al. 1982; de Vries et al. 1987). When those rhythms are disrupted during gestation, this can have detrimental effects as indicated by shift work during pregnancy. This results in relatively small babies and animal experiments show that it may also lead to glucose intolerance and insulin resistance in newborns (Varcoe et al. 2011; Bonzini et al. 2011). A detailed overview of *in utero* development of circadian rhythms was previously published (Du Pré et al. 2014).

5.4.2 Oscillation of Core Clock Genes upon Differentiation of Embryonic Stem Cells

Previously, researchers have reported the absence of a clear functional core clock in pluripotent embryonic stem cells (ESCs). Via bioluminescent reporter systems and the analysis of expression levels of clock genes, no clear circadian rhythms could be identified. However, when pushing these stem cells to leave their pluripotent, proliferative state through the withdrawal of leukemia inhibitory factor (LIF), circadian rhythms slowly started to emerge. By day 28, clear and robust rhythms were observed in D28 differentiated embryoid bodies (EBs) (Yagita et al. 2010; Umemura et al. 2013). The same holds true when stem cells are differentiated in a more directed way towards the neuronal lineage by addition of retinoic acid (Yagita et al. 2010; Kowalska et al. 2010). When differentiated cells are reprogrammed towards induced pluripotent stem cells (iPSCs), they lose their established oscillatory clock gene expression pattern. However, upon re-differentiation, they acquire an active circadian clock again. This suggests that an inherent circadian clock is linked to the differentiation status of a cell, and that this clock might be established during early phases of development (Brown 2014).

Albeit not clearly oscillating, most clock genes are expressed in ESCs. However, expression levels are different in comparison to differentiated cells. *Per1*, *Per2* and *Clock* are expressed at a lower level, *Cry2* is expressed at a higher level, and *Bmal1* shows similar transcript levels (Umemura et al. 2014). Apart from being molecular oscillators, this implies a different function of clock genes in ESCs. A possible explanation for the lack of clock gene oscillation in ESCs, is proposed in a recent paper of the Yagita lab. They show that the decline of Importin α2, encoded by *Kpn2α,* plays a key role in the acquirement of a circadian clock in vitro. Importin α2 is a nuclear transporter and shuttles specific pluripotency factors, like OCT3/4, into the nucleus. Additionally, it keeps differentiation linked factors, like OCT6, in the cytoplasm. Through this, a pluripotent state is retained (Yasuhara et al. 2013). Importin α2 also keeps clock factors like PER1 and PER2 in the cytoplasm (Umemura et al. 2014). Therefore, the absence of a functional clock in murine ESCs might be accountable to the fact that PER2 stays in the cytoplasm in pluripotent cells. The start of circadian oscillation is proposed as the timed entry of PER2 into the nucleus (Yagita et al. 2000).

5.4.3 Role of the Circadian Clock in Cell Cycle

Not only the circadian clock system, but also the cell cycle is an important biological oscillator. Researchers have intensively studied whether both processes are related and what the possible link between them might be. The cell cycle is represented by four different consecutive phases that ultimately lead to the division of proliferative cells. The S-phase is the replicative phase, in which the DNA is being duplicated. Next, G2 is a growth phase in which the cell contains a fully doubled genome. This leads to the M- (or mitotic) phase, where the DNA is propagated to the newly formed cell through actual division. M-phase leads to G1, which is another growth phase after which the cell re-enters the S-phase. However, non-dividing (somatic) cells can escape the cell cycle at G1 and reside in G0. However, they can switch back to a proliferative state via re-entering the cell cycle. The transition between cell cycle phases is facilitated by cyclin dependent kinases (CDKs) and cyclins, establishing so called "cell cycle checkpoints".

In the group of F. Naef, the influence of the cell cycle on the circadian clock and *vice versa*, have been studied at a single cell level in NIH3T3 cells. In the absence of a synchronizing stimulus, cell division precedes the peak of *Reverbα* expression with 5 h, concluding a clear 1:1 coupling between the cell cycle and the circadian clock (Bieler et al. 2014). This process is termed "Phase locking". When disturbing the cell cycle length, the circadian period is impacted but the 1:1 coupling remains. In reverse, lengthening the circadian period is not affecting the cell cycle, concluding a clear unidirectional coupling between cell cycle and the circadian clock (Bieler et al. 2014). On the other hand, cells in organs are under influence of synchronizing stimuli (e.g. light, temperature, hormones, feeding). Therefore, Feillet et al. found that synchronizing in vitro cultured cells leads to the occurrence of two subpopulations in culture: one in which the 1:1 phase locking is sustained, and another in which two circadian clock periods coincide with three complete cell cycles (Feillet et al. 2014). This suggests that the cell cycle is synchronized via physiological stimuli with the circadian clock (Feillet et al. 2015). Both studies indicate a clear interaction between cell cycle and the circadian clock, with the dominating factor being dependent on the environment (Feillet et al. 2015).

At a molecular level, the circadian clock has been linked to the cell cycle in several ways. The two main cell cycle phase windows controlled by clock genes can be found at the G2/M and the G1/S phase transition. Cell cycle regulators such as *Wee1* in the murine liver (Matsuo et al. 2003), *p20* and *p21* in a developing zebrafish embryo (Laranjeiro et al. 2013), *c-Myc* and *cyclin D1* (Fu et al. 2002) are under control of several clock proteins and show circadian rhythmicity in their expression.

It has been shown that the key component of the negative limb of the clock pathway, PER2, can regulate the cell cycle via inhibiting c-MYC activity. C-MYC directly blocks cyclin D1 activity, which is a roadblock for G1/S phase transition in the cell cycle. Overexpression of *Per2* concomitantly leads to cell cycle arrest in certain cancer cell lines and pushes these cells into apoptosis (Hua et al. 2006; Oda et al. 2009; Sun et al. 2010). Recently, a relatively new clock player termed NONO has also been implicated in cell cycle regulation. This multifunctional nuclear protein partners with PER2 and regulates expression of the cell cycle checkpoint gene *p16-Ink4A* in a circadian fashion. Since p16-INK4A facilitates G1 exit, NONO couples cell cycle to the circadian clock, which is argued to be a way to segregate cell proliferation from tissue organization in a time-based manner (Kowalska et al. 2013).

A growing amount of evidence links the circadian clock to the cell cycle and *vice versa*. A detailed description of the specific core players in this cell cycle/circadian clock network can be read in the following reviews (Kelleher et al. 2014; Feillet et al. 2015).

5.5 Circadian Rhythms in Adult Stem Cells

5.5.1 Regulation of Adult Stem Cells by Oscillatory Systems

Adult stem cells are multipotent stem cells that are present in adult organisms. They can proliferate and differentiate, but their potential is limited to specific cell types. Hematopoietic stem cells for example can differentiate into multiple blood cells, but not into skin cells. Adult stem cells are present in many organs. In some tissues they supply a constant renewal of cells (intestines, skin, blood), whereas in others they only become active after injury (heart, skeletal muscle). In the heart, adult stem cells of cardiac and non-cardiac origin are being investigated for their use in regenerative therapy.

24-hour rhythmicity plays an important role in adult stem cells. First, there are 24-h oscillations in stem cell mobilization and trafficking (Fig. 5.2). Hematopoietic stem cells (HSCs) in the circulation for example, show 24-h oscillations in both humans and mice (Laerum 1995; Méndez-Ferrer et al. 2008). In humans, there is a HSC release peak 5 h after the day-to-night transition. This rhythm is orchestrated by the circadian central clock, via diurnal noradrenalin secretion and local, sympathetic nerves to the bone marrow (Méndez-Ferrer et al. 2008). Oscillations in the number of circulating HSCs lead to 24-h oscillations in hematopoietic growth factors and the number of blood cells in the circulation (Laerum 1995; Gimble et al. 2009). As a consequence, 24-h oscillations are present in processes mediated by blood cells: diurnal rhythms in leukocytes for example cause oscillations in immune responses and rhythms in thrombocytes causes rhythms in coagulation time. Almost all adults have 24-h oscillations in HSC release. In aging and some diseases, however, these oscillations are dampened (Sletvold and Laerum 1988). Other adult stem cells in the circulation, such as endothelial progenitor cells, also show 24-h rhythms (Thomas et al. 2008).

5.5.2 Circadian Clock Gene Expression in Adult Stem Cells

Adult stem cells themselves also have molecular circadian clocks. Mesenchymal stem cells (MSCs), adult stem cells that can differentiate into cells of the mesoderm, such as osteoblasts, chondrocytes, and adipocytes, provide a good example of this. When MSCs are cultured in vitro, 24-h oscillations in gene expression of core clock genes such as *Bmal1*, *Cry1*, *Per2/3*, *Rev-erbα/β*, and *Dbp* are found, indicating that the peripheral circadian clock is present in these cells, independent of the presence of the central clock (Wu et al. 2008). Differences in expression between MSCs of old and young animals are observed (Yu et al. 2011). These rhythms seem to play an important role in physiology and pathophysiology. RORα for example, a component of the circadian clock present in MSCs, influences MSC differentiation (Meyer et al. 2000). When RORα is disrupted, bone mass parameters and bone geometry are impaired. Other studies have demonstrated that the circadian clock regulates bone metabolism, one of the functions of MSCs. Mice with altered circadian rhythms display bone remodeling and more than 26 % of the bone transcriptome exhibits circadian rhythmicity (Fu et al. 2005; Zvonic et al. 2007). Other functions of MSCs, such as adipose tissue homeostasis, also showed to be under circadian control (Guo et al. 2012).

Circadian rhythms in adult stem cells are associated with proliferation and differentiation (Fig. 5.2). Proliferation and differentiation are main characteristics of adult stem cells and vital for regenerative medicine. When MSCs are cultured in vitro and circadian rhythms are disrupted by translocation of CRY1 and PER1 from the nucleus to cytoplasm using laser irradiation, this leads to a change in differentiation from adipogenesis to osteogenesis (Kushibiki and Awazu 2008). In addition, genes that regulate stem cell proliferation and differentiation often show circadian rhythms in gene expression. There are genes, such as *GSK3* and *Sirt1*, which play an important role in both the circadian molecular clock and proliferation (Trowbridge et al. 2006; Yang et al. 2006). In addition, the phase of circadian rhythms seems to play a role in differentiation. In human epidermal stem cells the phase of circadian rhythms determines proliferation predisposition (Janich et al. 2013). When differentiation is induced by TGFβ or calcium, the core clock gene phase determines whether the epidermal stem cells respond to these cues. Based on circadian clock phase, stem cells can be divided into 'dormant' or 'active' stem cells (Janich et al. 2011). Disruption of clock genes such as *Bmal1* or *Per1/2* leads to accumulation or depletion of dormant stem cells.

In tissues that do not face a significant amount of proliferation during homeostatic conditions, such as the liver, heart, and skeletal muscle, circadian rhythms play a role in proliferation and differentiation after injury. After muscle injury for example, satellite cells from the basal lamina of muscle fibers become active to regenerate the muscle (Chatterjee et al. 2014). When the core clock gene *Bmal1* is disrupted, muscle regeneration is impaired because satellite cells cannot expand. In the liver, the circadian clock regulates cell cycle genes, which in turn regulate mitosis (Matsuo et al. 2003).

In summary, 24-h rhythms are important in the physiology of adult stem cells. The master clock orchestrates oscillations of adult stem cell numbers in the circulation, whereas peripheral clocks within stem cells regulate 24-h stem cell physiology. Specifically, circadian rhythms play an important role in proliferation and differentiation, both in normal physiology and after injury. Therefore, it would be interesting to determine if the dormancy/activity of cardiac stem cells is also clock mediated.

5.6 The Role of Circadian Rhythms in Tissue Regeneration for Cell-Based Therapy

The previous paragraphs have illustrated the importance of circadian rhythms in humans and specifically in the context of embryonic- and adult stem cells and their derivatives. Although research in this research field is scarce, it is likely that circadian rhythms will influence cell-based therapy.

First, circadian rhythms influence the number of adult stem cells present in different parts of the body. In (cardiac) regeneration therapy, sometimes stem cells from patients are collected and injected (Perin et al. 2014). When cells are harvested, their amount and quality could depend on the time of collection. Also, homing of the patient its own stem cells after injury, for example cardiac surgery, may be affected by 24-h rhythms. Depending on the time of injury, the body its regenerative capacity will differ, implying optimal regenerative time windows.

Second, the recipient body has 24-h rhythms in functions that are important in cell-based therapy. The immune response for example differs between day and night as seen in the pulmonary epithelium (Gibbs et al. 2014), but other important parameters fluctuate as well. Using cell-based therapy at the time-point that conditions in the recipient patient are best could benefit results.

Third, circadian rhythms of donor cells could be enhanced. Several types of stem cells have 24-h rhythms that influence their function. Injecting stem cells at a time-point optimal for the donor cells could also benefit regenerative therapies. This can be done by "resetting" cells in vitro to an optimal time-point using chemical substances (Izumo et al. 2006) or subjecting them to a serum shock (Balsalobre et al. 1998). This way, circadian rhythms between donor cells and recipient patient could be harmonized, possibly resulting in better engraftment of the donor cells.

Finally, circadian rhythms could be modified to enhance regenerative therapy, for example by targeting the molecular clock or its downstream pathways. Many techniques are available for this purpose, engaging on different levels ranging from genetic engineering of the molecular clock to altering clock in- or output signals (Bray et al. 2008; Tong et al. 2015). The effects of these alterations differ significantly. Some modifications for example will only slightly influence circadian phase or amplitude, whereas others completely abolish rhythmicity, with major impact on function (Sack et al. 2000; Bray et al. 2008). Modifications can be made both in donor cells and in the recipient patient.

In summary, optimal use of the knowledge on circadian rhythms and potentially modifying circadian rhythms or clock components could enhance stem cell differentiation and the effect of stem cell based cardiac repair.

References

Adamovich Y, Aviram R, Asher G (2014) The emerging roles of lipids in circadian control. Biochim Biophys Acta 1851(8):1017–1025. doi:10.1016/j.bbalip.2014.11.013

Akashi M, Takumi T (2005) The orphan nuclear receptor RORalpha regulates circadian transcription of the mammalian core-clock Bmal1. Nat Struct Mol Biol 12:441–448. doi:10.1038/nsmb925

Akhtar RA, Reddy AB, Maywood ES et al (2002) Circadian cycling of the mouse liver transcriptome, as revealed by cDNA microarray, is driven by the suprachiasmatic nucleus. Curr Biol 12:540–550

Aschoff J (1983) Circadian control of body temperature. J Therm Biol 8:143–147. doi:10.1016/0306-4565(83)90094-3

Asher G, Gatfield D, Stratmann M et al (2008) SIRT1 regulates circadian clock gene expression through PER2 deacetylation. Cell 134:317–328. doi:10.1016/j.cell.2008.06.050

Balsalobre A, Damiola F, Schibler U (1998) A serum shock induces circadian gene expression in mammalian tissue culture cells. Cell 93:929–937

Bieler J, Cannavo R, Gustafson K et al (2014) Robust synchronization of coupled circadian and cell cycle oscillators in single mammalian cells. Mol Syst Biol 10:739. doi:10.15252/msb.20145218

Bonzini M, Palmer KT, Coggon D et al (2011) Shift work and pregnancy outcomes: a systematic review with meta-analysis of currently available epidemiological studies. BJOG 118:1429–1437. doi:10.1111/j.1471-0528.2011.03066.x

Bordone L, Guarente L (2005) Calorie restriction, SIRT1 and metabolism: understanding longevity. Nat Rev Mol Cell Biol 6:298–305. doi:10.1038/nrm1616

Bray MS, Shaw CA, Moore MWS et al (2008) Disruption of the circadian clock within the cardiomyocyte influences myocardial contractile function, metabolism, and gene expression. Am J Physiol Heart Circ Physiol 294:H1036–H1047. doi:10.1152/ajpheart.01291.2007

Brown SA (2011) Circadian rhythms. A new histone code for clocks? Science 333:1833–1834. doi:10.1126/science.1212842

Brown SA (2014) Circadian clock-mediated control of stem cell division and differentiation: beyond night and day. Development 141:3105–3111. doi:10.1242/dev.104851

Brown SA, Azzi A (2013) Peripheral circadian oscillators in mammals. Handb Exp Pharmacol (217):45–66. doi: 10.1007/978-3-642-25950-0_3

Brown SA, Ripperger J, Kadener S et al (2005) PERIOD1-associated proteins modulate the negative limb of the mammalian circadian oscillator. Science 308:693–696. doi:10.1126/science.1107373

Bunger MK, Wilsbacher LD, Moran SM et al (2000) Mop3 is an essential component of the master circadian pacemaker in mammals. Cell 103:1009–1017. doi:10.1016/j.cell.2014.11.017

Cajochen C, Kräuchi K, Wirz-Justice A (2003) Role of melatonin in the regulation of human circadian rhythms and sleep. J Neuroendocrinol 15:432–437

Camacho F, Cilio M, Guo Y et al (2001) Human casein kinase Idelta phosphorylation of human circadian clock proteins period 1 and 2. FEBS Lett 489:159–165

Chatterjee S, Yin H, Nam D et al (2014) Brain and muscle Arnt-like 1 promotes skeletal muscle regeneration through satellite cell expansion. Exp Cell Res 331(1):200–210. doi:10.1016/j.yexcr.2014.08.041

Chen R, Schirmer A, Lee Y et al (2009) Rhythmic PER abundance defines a critical nodal point for negative feedback within the circadian clock mechanism. Mol Cell 36:417–430. doi:10.1016/j.molcel.2009.10.012

Chen R, D'Alessandro M, Lee C (2013) miRNAs are required for generating a time delay critical for the circadian oscillator. Curr Biol 23:1959–1968. doi:10.1016/j.cub.2013.08.005

Damiola F, Le Minh N, Preitner N et al (2000) Restricted feeding uncouples circadian oscillators in peripheral tissues from the central pacemaker in the suprachiasmatic nucleus. Genes Dev 14:2950–2961. doi:10.1101/gad.183500

Dardente H, Mendoza J, Fustin J-M et al (2008) Implication of the F-Box Protein FBXL21 in circadian pacemaker function in mammals. PLoS One 3, e3530. doi:10.1371/journal.pone.0003530

de Vries JI, Visser GH, Mulder EJ, Prechtl HF (1987) Diurnal and other variations in fetal movement and heart rate patterns at 20–22 weeks. Early Hum Dev 15:333–348

DeBruyne JP, Weaver DR, Reppert SM (2007) CLOCK and NPAS2 have overlapping roles in the suprachiasmatic circadian clock. Nat Neurosci 10:543–545. doi:10.1038/nn1884

Dierickx P, Doevendans PA, Geijsen N, van Laake LW (2012) Embryonic template-based generation and purification of pluripotent stem cell-derived cardiomyocytes for heart repair. J Cardiovasc Transl Res 5:566–580. doi:10.1007/s12265-012-9391-6

DiTacchio L, Le HD, Vollmers C et al (2011) Histone lysine demethylase JARID1a activates CLOCK-BMAL1 and influences the circadian clock. Science 333:1881–1885. doi:10.1126/science.1206022

Doi M, Hirayama J, Sassone-Corsi P (2006) Circadian regulator CLOCK is a histone acetyltransferase. Cell 125:497–508. doi:10.1016/j.cell.2006.03.033

Du Pré BC, van Veen TAB, Young ME et al (2014) Circadian rhythms in cell maturation. Physiology (Bethesda) 29:72–83. doi:10.1152/physiol.00036.2013
Du N-H, Arpat AB, De Matos M et al (2014) MicroRNAs shape circadian hepatic gene expression on a transcriptome-wide scale. Elife 3, e02510. doi:10.7554/eLife.02510
Dudley CA, Erbel-Sieler C, Estill SJ et al (2003) Altered patterns of sleep and behavioral adaptability in NPAS2-deficient mice. Science 301:379–383. doi:10.1126/science.1082795
Duffield GE, Best JD, Meurers BH et al (2002) Circadian programs of transcriptional activation, signaling, and protein turnover revealed by microarray analysis of mammalian cells. Curr Biol 12:551–557
Durgan DJ, Young ME (2010) The cardiomyocyte circadian clock: emerging roles in health and disease. Circ Res 106:647–658. doi:10.1161/CIRCRESAHA.109.209957
Durgan DJ, Pulinilkunnil T, Villegas-Montoya C et al (2010) Short communication: ischemia/reperfusion tolerance is time-of-day-dependent: mediation by the cardiomyocyte circadian clock. Circ Res 106:546–550. doi:10.1161/CIRCRESAHA.109.209346
Eide EJ, Woolf MF, Kang H et al (2005) Control of mammalian circadian rhythm by CKIepsilon-regulated proteasome-mediated PER2 degradation. Mol Cell Biol 25:2795–2807. doi:10.1128/MCB.25.7.2795-2807.2005
Etchegaray J-P, Lee C, Wade PA, Reppert SM (2003) Rhythmic histone acetylation underlies transcription in the mammalian circadian clock. Nature 421:177–182. doi:10.1038/nature01314
Falvey E, Marcacci L, Schibler U (1996) DNA-binding specificity of PAR and C/EBP leucine zipper proteins: a single amino acid substitution in the C/EBP DNA-binding domain confers PAR-like specificity to C/EBP. Biol Chem 377:797–809
Feillet C, Krusche P, Tamanini F et al (2014) Phase locking and multiple oscillating attractors for the coupled mammalian clock and cell cycle. Proc Natl Acad Sci U S A 111:9828–9833. doi:10.1073/pnas.1320474111
Feillet C, van der Horst GTJ, Levi F et al (2015) Coupling between the circadian clock and cell cycle oscillators: implication for healthy cells and malignant growth. Front Neurol 6:96. doi:10.3389/fneur.2015.00096
Fu L, Pelicano H, Liu J et al (2002) The circadian gene Period2 plays an important role in tumor suppression and DNA damage response in vivo. Cell 111:41–50
Fu L, Patel MS, Bradley A et al (2005) The molecular clock mediates leptin-regulated bone formation. Cell 122:803–815. doi:10.1016/j.cell.2005.06.028
Fustin J-M, Doi M, Yamaguchi Y et al (2013) RNA-methylation-dependent RNA processing controls the speed of the circadian clock. Cell 155:793–806. doi:10.1016/j.cell.2013.10.026
Gallego M, Kang H, Virshup DM (2006) Protein phosphatase 1 regulates the stability of the circadian protein PER2. Biochem J 399:169–175. doi:10.1042/BJ20060678
Gekakis N, Staknis D, Nguyen HB et al (1998) Role of the CLOCK protein in the mammalian circadian mechanism. Science 280:1564–1569
Gibbs J, Ince L, Matthews L et al (2014) An epithelial circadian clock controls pulmonary inflammation and glucocorticoid action. Nat Med 20:919–926. doi:10.1038/nm.3599
Gimble JM, Floyd ZE, Bunnell BA (2009) The 4th dimension and adult stem cells: can timing be everything? J Cell Biochem 107:569–578. doi:10.1002/jcb.22153
Griffin EA, Staknis D, Weitz CJ (1999) Light-independent role of CRY1 and CRY2 in the mammalian circadian clock. Science 286:768–771
Guillaumond F, Dardente H, Giguère V, Cermakian N (2005) Differential control of Bmal1 circadian transcription by REV-ERB and ROR nuclear receptors. J Biol Rhythms 20:391–403. doi:10.1177/0748730405277232
Guo B, Chatterjee S, Li L et al (2012) The clock gene, brain and muscle Arnt-like 1, regulates adipogenesis via Wnt signaling pathway. FASEB J 26:3453–3463. doi:10.1096/fj.12-205781
Harding HP, Lazar MA (1993) The orphan receptor Rev-ErbA alpha activates transcription via a novel response element. Mol Cell Biol 13:3113–3121
Hirano A, Yumimoto K, Tsunematsu R et al (2013) FBXL21 regulates oscillation of the circadian clock through ubiquitination and stabilization of cryptochromes. Cell 152:1106–1118. doi:10.1016/j.cell.2013.01.054

Hogenesch JB, Chan WK, Jackiw VH et al (1997) Characterization of a subset of the basic-helix-loop-helix-PAS superfamily that interacts with components of the dioxin signaling pathway. J Biol Chem 272:8581–8593

Hua H, Wang Y, Wan C et al (2006) Circadian gene mPer2 overexpression induces cancer cell apoptosis. Cancer Sci 97:589–596. doi:10.1111/j.1349-7006.2006.00225.x

Izumo M, Sato TR, Straume M, Johnson CH (2006) Quantitative analyses of circadian gene expression in mammalian cell cultures. PLoS Comput Biol 2, e136. doi:10.1371/journal.pcbi.0020136

Janich P, Pascual G, Merlos-Suárez A et al (2011) The circadian molecular clock creates epidermal stem …. Nature 480:209–214. doi:10.1038/nature10649

Janich P, Toufighi K, Solanas G et al (2013) Human epidermal stem cell function is regulated by circadian oscillations. Cell Stem Cell 13:745–753. doi:10.1016/j.stem.2013.09.004

Johnson MH, Lim A, Fernando D, Day ML (2002) Circadian clockwork genes are expressed in the reproductive tract and conceptus of the early pregnant mouse. Reprod Biomed Online 4:140–145

Katada S, Sassone-Corsi P (2010) The histone methyltransferase MLL1 permits the oscillation of circadian gene expression. Nat Struct Mol Biol 17:1414–1421. doi:10.1038/nsmb.1961

Keesler GA, Camacho F, Guo Y et al (2000) Phosphorylation and destabilization of human period I clock protein by human casein kinase I epsilon. Neuroreport 11:951–955

Kelleher FC, Rao A, Maguire A (2014) Circadian molecular clocks and cancer. Cancer Lett 342:9–18. doi:10.1016/j.canlet.2013.09.040

Kennaway DJ, Varcoe TJ, Voultsios A, Boden MJ (2013) Global loss of bmal1 expression alters adipose tissue hormones, gene expression and glucose metabolism. PLoS One 8, e65255. doi:10.1371/journal.pone.0065255

Kim JY, Kwak PB, Weitz CJ (2014) Specificity in circadian clock feedback from targeted reconstitution of the NuRD corepressor. Mol Cell 56:738–748. doi:10.1016/j.molcel.2014.10.017

Kojima S, Sher-Chen EL, Green CB (2012) Circadian control of mRNA polyadenylation dynamics regulates rhythmic protein expression. Genes Dev 26:2724–2736. doi:10.1101/gad.208306.112

Kowalska E, Moriggi E, Bauer C et al (2010) The circadian clock starts ticking at a developmentally early stage. J Biol Rhythms 25:442–449. doi:10.1177/0748730410385281

Kowalska E, Ripperger JA, Hoegger DC et al (2013) NONO couples the circadian clock to the cell cycle. Proc Natl Acad Sci U S A 110:1592–1599. doi:10.1073/pnas.1213317110

Kume K, Zylka MJ, Sriram S et al (1999) mCRY1 and mCRY2 are essential components of the negative limb of the circadian clock feedback loop. Cell 98:193–205

Kushibiki T, Awazu K (2008) Controlling osteogenesis and adipogenesis of mesenchymal stromal cells by regulating a circadian clock protein with laser irradiation. Int J Med Sci 5:319–326

Laerum OD (1995) Hematopoiesis occurs in rhythms. Exp Hematol 23:1145–1147. doi:10.1038/ncomms8056

Lamia KA, Storch K-F, Weitz CJ (2008) Physiological significance of a peripheral tissue circadian clock. Proc Natl Acad Sci U S A 105:15172–15177. doi:10.1073/pnas.0806717105

Laranjeiro R, Tamai TK, Peyric E et al (2013) Cyclin-dependent kinase inhibitor p20 controls circadian cell-cycle timing. Proc Natl Acad Sci U S A 110:6835–6840. doi:10.1073/pnas.1217912110

Lowrey PL, Shimomura K, Antoch MP et al (2000) Positional syntenic cloning and functional characterization of the mammalian circadian mutation tau. Science 288:483–492

Lumaban JG, Nelson DL (2014) The Fragile X proteins Fmrp and Fxr2p cooperate to regulate glucose metabolism in mice. Hum Mol Genet 24(8):2175–2184. doi:10.1093/hmg/ddu737

Marcheva B, Ramsey KM, Buhr ED et al (2010) Disruption of the clock components CLOCK and BMAL1 leads to hypoinsulinaemia and diabetes. Nature 466:627–631. doi:10.1038/nature09253

Matsuo T, Yamaguchi S, Mitsui S et al (2003) Control mechanism of the circadian clock for timing of cell division in vivo. Science 302:255–259. doi:10.1126/science.1086271

McGlincy NJ, Valomon A, Chesham JE et al (2012) Regulation of alternative splicing by the circadian clock and food related cues. Genome Biol 13:R54. doi:10.1186/gb-2012-13-6-r54

Méndez-Ferrer S, Lucas D, Battista M, Frenette PS (2008) Haematopoietic stem cell release is regulated by circadian oscillations. Nature 452:442–447. doi:10.1038/nature06685

Merrow M, Spoelstra K, Roenneberg T (2005) The circadian cycle: daily rhythms from behaviour to genes. EMBO Rep 6:930–935. doi:10.1038/sj.embor.7400541

Meyer T, Kneissel M, Mariani J, Fournier B (2000) In vitro and in vivo evidence for orphan nuclear receptor RORalpha function in bone metabolism. Proc Natl Acad Sci U S A 97:9197–9202. doi:10.1073/pnas.150246097

Milagro FI, Gómez-Abellán P, Campión J et al (2012) CLOCK, PER2 and BMAL1 DNA methylation: association with obesity and metabolic syndrome characteristics and monounsaturated fat intake. Chronobiol Int 29:1180–1194. doi:10.3109/07420528.2012.719967

Miller BH, McDearmon EL, Panda S et al (2007) Circadian and CLOCK-controlled regulation of the mouse transcriptome and cell proliferation. Proc Natl Acad Sci U S A 104:3342–3347. doi:10.1073/pnas.0611724104

Morf J, Rey G, Schneider K et al (2012) Cold-inducible RNA-binding protein modulates circadian gene expression posttranscriptionally. Science 338:379–383. doi:10.1126/science.1217726

Nagoshi E, Saini C, Bauer C et al (2004) Circadian gene expression in individual fibroblasts: cell-autonomous and self-sustained oscillators pass time to daughter cells. Cell 119:693–705. doi: 10.1016/j.cell.2004.11.015

Nakahata Y, Kaluzova M, Grimaldi B et al (2008) The NAD+-dependent deacetylase SIRT1 modulates CLOCK-mediated chromatin remodeling and circadian control. Cell 134:329–340. doi:10.1016/j.cell.2008.07.002

Nam HJ, Boo K, Kim D et al (2014) Phosphorylation of LSD1 by PKCα is crucial for circadian rhythmicity and phase resetting. Mol Cell 53:791–805. doi:10.1016/j.molcel.2014.01.028

Oda A, Katayose Y, YABUUCHI S et al (2009) Clock gene mouse period2 overexpression inhibits growth of human pancreatic cancer cells and has synergistic effect with cisplatin. Anticancer Res 29:1201–1209

Ohno T, Onishi Y, Ishida N (2007) A novel E4BP4 element drives circadian expression of mPeriod2. Nucleic Acids Res 35:648–655. doi:10.1093/nar/gkl868

Paschos GK, FitzGerald GA (2010) Circadian clocks and vascular function. Circ Res 106:833–841. doi:10.1161/CIRCRESAHA.109.211706

Perin EC, Sanz-Ruiz R, Sánchez PL et al (2014) Adipose-derived regenerative cells in patients with ischemic cardiomyopathy: The PRECISE Trial. Am Heart J 168:88–95.e2. doi:10.1016/j.ahj.2014.03.022

Preitner N, Damiola F, Lopez-Molina L et al (2002) The orphan nuclear receptor REV-ERBalpha controls circadian transcription within the positive limb of the mammalian circadian oscillator. Cell 110:251–260

Preußner M, Wilhelmi I, Schultz A-S et al (2014) Rhythmic U2af26 alternative splicing controls PERIOD1 stability and the circadian clock in mice. Mol Cell 54:651–662. doi:10.1016/j.molcel.2014.04.015

Reppert SM, Weaver DR (2002) Coordination of circadian timing in mammals. Nature 418(6901):935–941

Rudic RD, McNamara P, Curtis A-M et al (2004) BMAL1 and CLOCK, two essential components of the circadian clock, are involved in glucose homeostasis. PLoS Biol 2, e377. doi:10.1371/journal.pbio.0020377

Sack RL, Brandes RW, Kendall AR, Lewy AJ (2000) Entrainment of free-running circadian rhythms by melatonin in blind people. N Engl J Med 343:1070–1077. doi:10.1056/NEJM200010123431503

Sassone-Corsi P (2012) Minireview: NAD+, a circadian metabolite with an epigenetic twist. Endocrinology 153:1–5. doi:10.1210/en.2011-1535

Sato TK, Panda S, Miraglia LJ et al (2004) A functional genomics strategy reveals Rora as a component of the mammalian circadian clock. Neuron 43:527–537. doi:10.1016/j.neuron.2004.07.018

Saxena MT, Aton SJ, Hildebolt C et al (2007) Bioluminescence imaging of period1 gene expression in utero. Mol Imaging 6:68–72

Scheiermann C, Kunisaki Y, Frenette PS (2013) Circadian control of the immune system. Nat Rev Immunol 13:190–198. doi:10.1038/nri3386

Shi S, Hida A, McGuinness OP et al (2010) Circadian clock gene Bmal1 is not essential; functional replacement with its paralog, Bmal2. Curr Biol 20:316–321. doi:10.1016/j.cub.2009.12.034

Siepka SM, Yoo S-H, Park J et al (2007) Circadian mutant overtime reveals F-box protein FBXL3 regulation of cryptochrome and period gene expression. Cell 129:1011–1023. doi:10.1016/j.cell.2007.04.030

Sletvold O, Laerum OD (1988) Multipotent stem cell (CFU-S) numbers and circadian variations in aging mice. Eur J Haematol 41:230–236

Son GH, Chung S, Choe HK et al (2008) Adrenal peripheral clock controls the autonomous circadian rhythm of glucocorticoid by causing rhythmic steroid production. Proc Natl Acad Sci U S A 105:20970–20975. doi:10.1073/pnas.0806962106

Stokkan KA, Yamazaki S, Tei H et al (2001) Entrainment of the circadian clock in the liver by feeding. Science 291:490–493. doi:10.1126/science.291.5503.490

Sun C-M, Huang S-F, Zeng J-M et al (2010) Per2 inhibits k562 leukemia cell growth in vitro and in vivo through cell cycle arrest and apoptosis induction. Pathol Oncol Res 16:403–411. doi:10.1007/s12253-009-9227-0

Taegtmeyer H (2000) Metabolism—the lost child of cardiology. J Am Coll Cardiol 36:1386–1388

Thomas HE, Redgrave R, Cunnington MS et al (2008) Circulating endothelial progenitor cells exhibit diurnal variation. Arterioscler Thromb Vasc Biol 28:e21–e22. doi:10.1161/ATVBAHA.107.160317

Tong X, Zhang D, Arthurs B et al (2015) Palmitate inhibits SIRT1-dependent BMAL1/CLOCK interaction and disrupts circadian gene oscillations in hepatocytes. PLoS One 10, e0130047. doi:10.1371/journal.pone.0130047

Trowbridge JJ, Xenocostas A, Moon RT, Bhatia M (2006) Glycogen synthase kinase-3 is an in vivo regulator of hematopoietic stem cell repopulation. Nat Med 12:89–98. doi:10.1038/nm1339

Tsai J-Y, Kienesberger PC, Pulinilkunnil T et al (2010) Direct regulation of myocardial triglyceride metabolism by the cardiomyocyte circadian clock. J Biol Chem 285:2918–2929. doi:10.1074/jbc.M109.077800

Ueda HR, Chen W, Adachi A et al (2002) A transcription factor response element for gene expression during circadian night. Nature 418:534–539. doi:10.1038/nature00906

Umemura Y, Yoshida J, Wada M et al (2013) An in vitro ES cell-based clock recapitulation assay model identifies CK2α as an endogenous clock regulator. PLoS One 8, e67241. doi:10.1371/journal.pone.0067241

Umemura Y, Koike N, Matsumoto T et al (2014) Transcriptional program of Kpna2/Importin-α2 regulates cellular differentiation-coupled circadian clock development in mammalian cells. Proc Natl Acad Sci U S A 111(47):E5039–E5048. doi:10.1073/pnas.1419272111

Unsal-Kaçmaz K, Mullen TE, Kaufmann WK, Sancar A (2005) Coupling of human circadian and cell cycles by the timeless protein. Mol Cell Biol 25:3109–3116. doi: 10.1128/MCB.25.8.3109-3116.2005

van Der Horst GT, Muijtjens M, Kobayashi K et al (1999) Mammalian Cry1 and Cry2 are essential for maintenance of circadian rhythms. Nature 398:627–630. doi:10.1038/19323

Varcoe TJ, Wight N, Voultsios A et al (2011) Chronic phase shifts of the photoperiod throughout pregnancy programs glucose intolerance and insulin resistance in the rat. PLoS One 6, e18504. doi:10.1371/journal.pone.0018504

Visser GH, Goodman JD, Levine DH, Dawes GS (1982) Diurnal and other cyclic variations in human fetal heart rate near term. Am J Obstet Gynecol 142:535–544

Vitaterna MH, Selby CP, Todo T et al (1999) Differential regulation of mammalian period genes and circadian rhythmicity by cryptochromes 1 and 2. Proc Natl Acad Sci U S A 96:12114–12119

Wang C-Y, Wen M-S, Wang H-W et al (2008) Increased vascular senescence and impaired endothelial progenitor cell function mediated by mutation of circadian gene Per2. Circulation 118:2166–2173. doi:10.1161/CIRCULATIONAHA.108.790469
Westgate EJ, Cheng Y, Reilly DF et al (2008) Genetic components of the circadian clock regulate thrombogenesis in vivo. Circulation 117:2087–2095. doi:10.1161/CIRCULATIONAHA.107.739227
Wu X, Yu G, Parks H et al (2008) Circadian mechanisms in murine and human bone marrow mesenchymal stem cells following dexamethasone exposure. Bone 42:861–870. doi:10.1016/j.bone.2007.12.226
Yagita K, Yamaguchi S, Tamanini F et al (2000) Dimerization and nuclear entry of mPER proteins in mammalian cells. Genes Dev 14:1353–1363
Yagita K, Tamanini F, Yasuda M et al (2002) Nucleocytoplasmic shuttling and mCRY-dependent inhibition of ubiquitylation of the mPER2 clock protein. EMBO J 21:1301–1314. doi:10.1093/emboj/21.6.1301
Yagita K, Horie K, Koinuma S et al (2010) Development of the circadian oscillator during differentiation of mouse embryonic stem cells in vitro. Proc Natl Acad Sci U S A 107:3846–3851. doi:10.1073/pnas.0913256107
Yang X, Downes M, Yu RT et al (2006) Nuclear receptor expression links the circadian clock to metabolism. Cell 126:801–810. doi:10.1016/j.cell.2006.06.050
Yasuhara N, Yamagishi R, Arai Y et al (2013) Importin alpha subtypes determine differential transcription factor localization in embryonic stem cells maintenance. Dev Cell 26:123–135. doi:10.1016/j.devcel.2013.06.022
Yoo S-H, Ko CH, Lowrey PL et al (2005) A noncanonical E-box enhancer drives mouse Period2 circadian oscillations in vivo. Proc Natl Acad Sci U S A 102:2608–2613. doi:10.1073/pnas.0409763102
Yoo S-H, Mohawk JA, Siepka SM et al (2013) Competing E3 ubiquitin ligases govern circadian periodicity by degradation of CRY in nucleus and cytoplasm. Cell 152:1091–1105. doi:10.1016/j.cell.2013.01.055
Yu JM, Wu X, Gimble JM et al (2011) Age-related changes in mesenchymal stem cells derived from rhesus macaque bone marrow. Aging Cell 10:66–79. doi:10.1111/j.1474-9726.2010.00646.x
Yu X, Rollins D, Ruhn KA et al (2013) TH17 cell differentiation is regulated by the circadian clock. Science 342:727–730. doi:10.1126/science.1243884
Zvonic S, Ptitsyn AA, Kilroy G et al (2007) Circadian oscillation of gene expression in murine calvarial bone. J Bone Miner Res 22:357–365. doi:10.1359/jbmr.061114

Chapter 6
Cardiac Regeneration and microRNAs: Regulators of Pluripotency, Reprogramming, and Cardiovascular Lineage Commitment

Martijn J.H. Doeleman, Dries A.M. Feyen, Christina F. de Veij Mestdagh, and Joost P.G. Sluijter

microRNAs (miRNAs), ~22 nucleotide noncoding RNAs, are post-transcriptional regulators of gene expression that are involved in the regulation of almost every biological process, including self-renewal, pluripotency, and differentiation (Tüfekci et al. 2014). By inhibiting mRNA translation or inducing mRNA degradation, these small noncoding RNA molecules modulate gene expression. There are almost 2000 miRNAs encoded in the human genome. Each miRNA can target up to several hundred complementary mRNAs, giving miRNAs the ability to control gene expression patterns rather than single genes (Friedman et al. 2009). miRNA expression profiling studies have demonstrated that each cell type possesses a unique miRNA expression pattern, which results in different cellular characteristics (Liang et al. 2007). Because miRNAs control gene expression networks, they might represent a useful strategy to coordinate proliferation and differentiation of cells that are able to contribute to cardiac repair and regeneration.

This chapter will describe the mechanism of miRNA-mediated gene silencing and the role of miRNAs in the regulation of different processes, including pluripotency and self-renewal of stem cells, reprogramming of somatic cells, and differentiation of stem cells and progenitor cells into cardiovascular cell types. Altogether, this chapter will give insight into the potential of miRNAs for cardiac regeneration.

M.J.H. Doeleman • D.A.M. Feyen • C.F. de Veij Mestdagh
Division Heart and Lung, Cardiology, Experimental Cardiology Laboratory, University Medical Center Utrecht, Room number G03.550, P.O. Box 85500, Utrecht 3508 GA, The Netherlands

J.P.G. Sluijter, Ph.D. (✉)
Division Heart and Lung, Cardiology, Experimental Cardiology Laboratory, University Medical Center Utrecht, Room number G03.550, P.O. Box 85500, Utrecht 3508 GA, The Netherlands

ICIN, Netherlands Heart Institute, Catharijnesingel 52, Utrecht 3511 GC, The Netherlands
e-mail: j.sluijter@umcutrecht.nl

R. Madonna (ed.), *Stem Cells and Cardiac Regeneration*, Stem Cell Biology and Regenerative Medicine, DOI 10.1007/978-3-319-25427-2_6

6.1 The Cellular Function of microRNAs

6.1.1 *Introduction*

The discovery of microRNAs (miRNAs) led to a radical change in the field of RNA molecular biology. Until 1993, there was no knowledge of one of the main mechanisms of gene regulation. Today, miRNAs are acknowledged as major regulators in many physiological processes (Tüfekci et al. 2014). The growing awareness of the importance of miRNAs in different biological processes has been followed by an explosion of scientific publications. In these publications, it has been estimated that over 60 % of the human protein-coding genes are regulated by miRNAs (Friedman et al. 2009). miRNAs were first discovered by Victor Ambros and colleagues who performed a genetic screen to investigate defects during *Caenorhabditis elegans* development. They discovered that the gene *lin-4,* a repressor of *lin-14*, did not encode a protein. Instead, *lin-4* encoded a pair of small RNAs of ~22 and ~61 nucleotides in length. The longer RNA was proposed to be the precursor of the shorter RNA molecule. Moreover, *lin-4* RNA exhibited antisense complementarity to sites in the 3′-untranslated region (UTR) of the *lin-14* gene. This suggested an antisense regulatory mechanism (Lee et al. 1993). The shorter *lin-4* RNA is now recognized as the first member of an elaborate group of small RNAs involved in gene suppression (Lee and Ambros 2001). However, before the turn of the century, there was no evidence for other small non-coding RNAs similar to *lin-4*. Fortunately, the discovery of Let-7 gave insight into the big world of these small RNAs. Let-7 encodes a 21 nucleotide small RNA involved in the transition to the adult stage of *C. elegans* (Reinhart et al. 2000), and also inhibits translation of its target by binding to the 3'-UTR (Abrahante et al. 2003). In contrast to *lin-4*, orthologs of Let-7 were found in different species, including molluscs, sea urchins, flies and humans (Hertel et al. 2012). This suggested a general and widespread role for small RNAs in the regulation of gene expression during development. Further studies have shown that small RNAs perform a regulatory function in almost every biological process (Tüfekci et al. 2014). In 2001, Lagos-Quintana et al. proposed the term 'microRNAs' for this still growing family of small RNAs (Lagos-Quintana et al. 2001). Currently, the latest version of miRBase (mirbase.org) contains 1881 mature human miRNAs (Kozomara and Griffiths-Jones 2014). Therefore, miRNAs compose one of the largest families of gene regulatory molecules in the human genome.

6.1.2 *Biogenesis of microRNAs*

In the most basic definition, miRNAs are small non-coding RNAs of approximately 22 nucleotides long, which control gene expression by inhibiting messenger RNA (mRNA) translation and inducing mRNA instability and degradation (Huntzinger and Izaurralde 2011). Computational predictions of microRNA targets estimate that each human miRNA can control up to hundreds of different mRNAs (Xu et al. 2014). In the human genome, approximately one third of the miRNAs have their own promoter. Other microRNAs are found in exonic or intronic regions of either non-coding or coding genes (Saini et al. 2007). The majority of intronic microRNAs are transcribed from the same promoter as their host gene. However, approximately one third of the intronic miRNAs are transcribed from an independent promoter, allowing for more regulation of their transcription (Ozsolak et al. 2008).

miRNA biogenesis starts with the synthesis of a primary miRNA (pri-miRNA) transcript, consisting of up to several thousand nucleotides that contain stem-loop structures and flanking single strand segments. Pri-miRNAs are largely transcribed by RNA polymerase II, as capped and polyadenylated transcripts (Cai et al. 2004). Pri-miRNAs can contain one stem loop structure (pre-miRNA) or a cluster of several stem loop structures. Pri-miRNAs are processed by the microprocessor-complex into –70 nucleotide pre-miRNA within the nucleus. The major components of the microprocessor-complex are the RNase type III endonuclease Drosha and cofactor Di George syndrome critical region 8 (DGCR8). DGCR8 interacts with the pri-miRNA at the junction of the stem loop and the single strand segments, and directs Drosha to cleave approximately 11 base pairs (bp) downstream of this junction (Han et al. 2006). In turn, Drosha cleaves the hairpin secondary structure out of the pri-miRNA, generating pre-miRNA with 3′ overhangs (Han et al. 2004).

After pri-miRNA processing by the microprocessor, the resulting ~70 nucleotide hairpin pre-miRNA is transported out of the nucleus by Exportin 5, a RAN-GTP dependent nucleo/cytoplasmatic cargo transporter (Lund et al. 2004). A small subclass of miRNAs, located in intronic regions (mirtrons), can bypass processing by the microprocessor and are directly transported to the cytoplasm by Exportin 5 (Westholm and Lai 2011). In the cytoplasm, the pre-miRNA is again processed by an RNase type III enzyme, called Dicer, which recognizes the 3′ overhang of the pre-miRNA and cleaves within the stem loop, generating an imperfect ~22 nucleotide miRNA-duplex. Dicer associates with dsRNA binding domain proteins, transactivating response RNA-binding protein (TRBP) and protein activator of PKR (PACT), which assist with Dicer's cleavage precision, facilitating accurate cleavage of miRNA duplexes (Wilson et al. 2014). Subsequently, this complex recruits one of the Argonaute proteins (AGO1-4) and a GW182 protein (Pfaff et al. 2013), which form the key components of the multi-protein miRNA-induced silencing complex (miRISC) (Chendrimada et al. 2005). The function of the miRISC is to select and recruit one of the strands of the miRNA duplex, which guides the catalytic complex to its complementary mRNA. In result, the miRISC induces inhibition of translation and/or instability and subsequent degradation of the mRNA (Huntzinger and Izaurralde 2011). Fig. 6.1 displays the biogenesis of miRNAs.

6.1.3 microRNA Target Recognition

The miRNA within the miRISC targets mRNA molecules to induce miRNA-mediated translational repression as well as miRNA-mediated mRNA decay. Target interaction between mRNAs and miRNAs involves seed-pairing between nucleotides of both

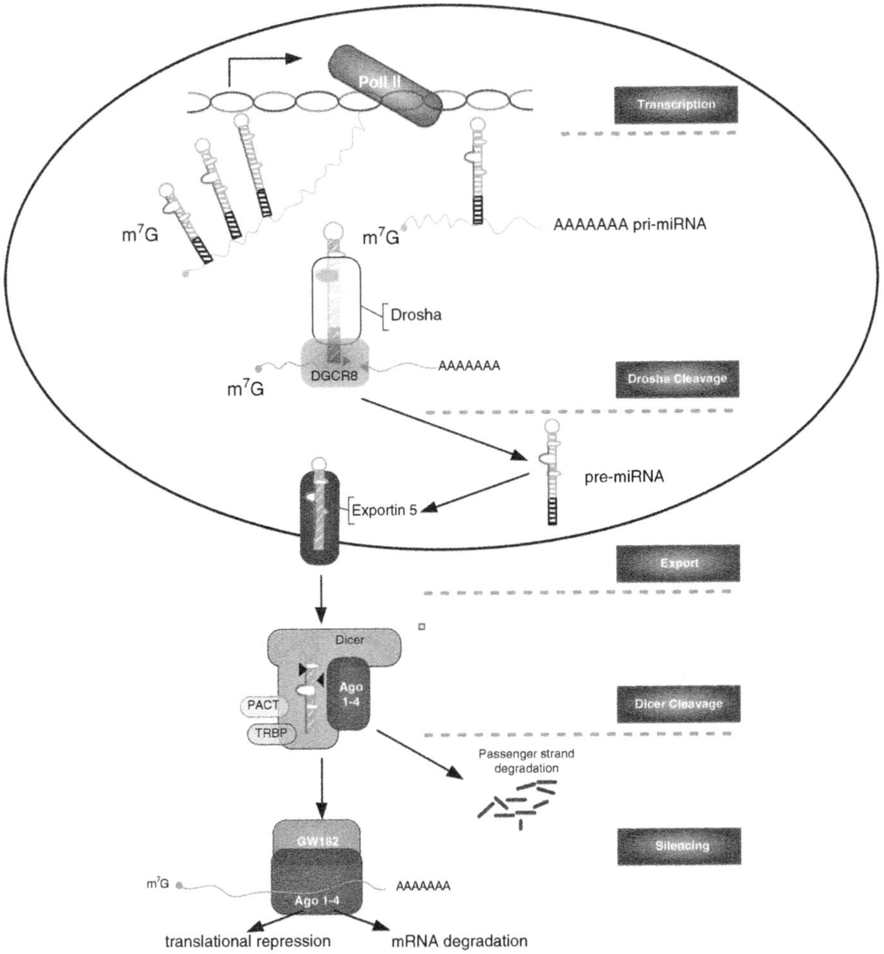

Fig. 6.1 The miRNA biogenesis pathway. In the nucleus, pri-miRNAs are transcribed by RNA polymerase II as capped (m7G, 7-methylguanosine-cap) and polyadenylated transcripts. The pri-miRNA can consist of one pre-miRNA or a cluster of pre-miRNAs. Processing of the pri-miRNA is catalyzed by the micro-processor complex (Drosha and DGCR8) inside the nucleus. Following micro-processing, the pre-miRNA is transported out of the nucleus by RAN-GTP dependent transporter Exportin 5. In the cytoplasm, the pre-miRNA is processed by Dicer, assisted by TRBP and PACT. The generated miRNA duplex is loaded into the miRISC, composed of an Ago protein (AGO1-4) and GW182. This is followed by guide strand selection. The mature miRNA inside the miRISC (*red*) guides the complex to the target mRNA

strands. Recognition of the mRNA target does not require perfect complementarity. In mammals, mRNA target recognition generally involves mRNA sites in the 3′-UTR that base-pair with approximately 7 nucleotides near the 5′-end of the miRNA. These miRNA nucleotides are termed 'the seed sequence' (Lewis et al. 2005). However, miRNA target sites in the coding sequence of mRNAs have also been discovered and miRNA target sites that cannot be explained by this canonical seed-pairing model also exist. This makes miRNA target identification a very difficult task (Thomas et al. 2010). Even though the miRNA guides the miRISC, the protein components of the miRISC execute the silencing of target mRNAs.

6.1.4 microRNA-Induced Silencing Complex

The first miRNA, *lin-4*, was reported to inhibit translation of *lin-14* mRNA without destabilizing the mRNA molecule (Lee et al. 1993). There have been several studies using *C. elegans* and mammalian cell cultures that only observed translational repression at the initiation of translation (Ding and Grosshans 2009; Bhattacharyya et al. 2006). In contrast, others reported miRNA-mediated deadenylation, decapping, and subsequent decay of miRNA-target molecules (Eulalio et al. 2009). Over the past few years, there has been growing evidence for translational repression as well as mRNA decay by miRNAs (Huntzinger and Izaurralde 2011).

Recent studies on translational repression of miRNAs propose that the miRISC interferes with the function of the cap-binding complex. Mammalian mRNAs exist as capped and polyadenylated transcripts that are circularized by protein complexes interacting with the 5′-cap structure (comprising the cap-binding complex) and the 3′-poly(A)-tail. Circularization of mRNAs is necessary for efficient translation and interfering with the cap-binding complex can result in loss of circularization and subsequent translational repression (Mathonnet et al. 2007; Fukao et al. 2014). In addition, GW182 proteins are known to interact with poly(A) binding proteins (PABP). Recently, it was found that interaction between GW182 and PABP leads to disassociation of PABP from the mRNA molecule. This can cause disruption of mRNA circularization and inhibit efficient translation, hereby facilitating translational repression (Zekri et al. 2013).

A second mechanism by which miRNAs regulate target mRNAs is mRNA destabilization and subsequent mRNA decay. This process is initiated by deadenylation of the mRNA molecule by one of the deadenylase complexes recruited by GW182. GW182 interacts directly with two deadenylase complexes, CCR4-NOT and, to a lesser extent, PAN2-PAN3 (Braun et al. 2011). The deadenylase complexes produce an unstable mRNA molecule without a poly(A)-tail. Subsequently, the miRISC proteins initiate removal of the 5′-cap (m^7G) of the target mRNA. This process is called decapping and requires activators mRNA-decapping enzyme 1 (DCP1), Me31B, and HPat, which are recruited by the miRISC. These factors attract mRNA-decapping enzyme 2 (DCP2), which catalyzes decapping of the target mRNA (Nishihara et al. 2013). This is an irreversible process that commits the mRNA to full degradation by the major cytoplasmic 5′–3′ exonuclease XRN1 (Braun et al. 2012).

Studies on the relationship between translational repression and mRNA decay have demonstrated that translational repression contributed little to the repression of endogenous mRNAs in comparison with mRNA decay (10–25 % and 66–90 %, respectively) (Eichhorn et al. 2014; Guo et al. 2010; Hendrickson et al. 2009). However, these global measurements of translational efficiency and mRNA levels were often made at relatively late time points after introducing miRNAs. Therefore, they are thought to reflect the long-term steady-state effects of miRNAs. Two recent studies examined the initial effect of miRNAs on inducible reporter genes. They concluded that miRNAs predominantly exert their effects through translational repression on newly synthesized targets. This step is followed by mRNA deadenylation, decapping and subsequent mRNA decay, which is the dominant effect of miRNAs at steady state conditions (Djuranovic et al. 2012; Béthune et al. 2012).

Abovementioned mechanism enables miRNAs to regulate gene expression at a post-transcriptional level. As earlier mentioned, miRNAs are differentially expressed in different cells and cell types, including embryonic stem cells (ESCs), induced pluripotent stem (iPS) cells, and fully differentiated cells, resulting in different gene expression profiles and cellular properties (Wilson et al. 2009; Houbaviy et al. 2003). Many studies have examined genome-wide miRNA expression profiles of stem cells during differentiation or self-renewal to improve understanding of miRNAs involved in the regulatory networks of these cellular processes (Mallon et al. 2014; Razak et al. 2013).

6.2 The Role of microRNAs in Pluripotent Stem Cells

6.2.1 Introduction

Stem cells are undifferentiated cells that are able renew themselves, and differentiate into specialized cells to regenerate tissues. The potency of stem cells describes the potential to differentiate into different cell types. Totipotent stem cells can differentiate into every embryonic and extra embryonic cell type, and are able to generate an entire organism. However, only cells produced by the first few divisions of a fertilized egg are totipotent. Pluripotent cells are descendants of totipotent cells and can differentiate into cells of all embryonic cell types, but not extra embryonic cell types. Furthermore, adult multipotent stem cells, often termed progenitor cells, can only differentiate into several closely related cell types (Mitalipov and Wolf 2009). Pluripotent stem cells hold significant potential for clinical therapies because they theoretically possess the capacity to regenerate all types of tissue. However, these characteristics are only displayed by cells of the inner cell mass (ICM) from early embryos. Fortunately, early studies demonstrated that the pluripotent state of stem cells can be captured by placing pre-implantation blastocysts in culture, leading to the generation of pluripotent embryonic stem cells (ESCs) (Evans and Kaufman 1981).

Furthermore, the derivation of complete adult and sexually mature animals by transplantation of somatic nuclei into enucleated oöcytes demonstrated that somatic nuclei also withhold the information to generate a complete adult organism (Gurdon et al. 1958). Further studies demonstrated that differentiated cells can also be reprogrammed to pluripotent stem cells by nuclear transfer or cell fusion (Campbell et al. 1996; Blau et al. 1983). A real breakthrough in the reprogramming of somatic cells was made when Shinya Yamanaka invented the induced pluripotent stem (iPS) cell technology in 2006. Yamanaka and co-workers demonstrated that pluripotency can be induced by introducing four transcription factors, Oct4, Sox2, Klf4, and c-Myc (OSKM), in mouse adult fibroblasts by retroviral transfection (Takahashi and Yamanaka 2006). This approach offers the opportunity to generate pluripotent stem cells from differentiated cells of an individual, which opened the field of personalized regenerative medicine. Unfortunately, reprogramming with OSKM has remained slow (2–4 weeks) and inefficient (0.01–0.1 % of total cells) (Stadtfeld and Hochedlinger 2010).

In the last few years, a variety of pluripotent and adult stem cells have been proposed to promote cardiac regeneration. The heart itself possesses little regenerative capacity, since cardiomyocytes exhibit a very slow turnover and the heart lacks a sufficient stem cell reservoir to regenerate itself. In this regard, pluripotent stem cells and progenitor cells have been proposed to replenish the loss of cardiomyocytes or to induce proliferation and differentiation of resident stem cells in the heart. Several subsets of stem cells and cardiac progenitor cells have shown experimental and clinical benefits. Alas, the effects of cell-based therapy have been modest, often due to low engraftment of applied cells and the limited differentiation into cardiovascular cell types of adult progenitor cells (Li et al. 2012; Balsam et al. 2004). In contrast, ESCs and iPS cells are able to differentiate into all cardiovascular lineages. However, efficient differentiation before transplantation is necessary since pluripotent cells may lead to teratomas after delivery in patients (Sun et al. 2010).

Because miRNAs are powerful regulators that control gene expression networks, they might represent useful tools to promote reprogramming of somatic cells into iPS cells or direct somatic cells to specific lineages in vivo and in vitro. Understanding miRNAs involved in self-renewal, reprogramming and differentiation may enable the development of safer and more efficient cell therapies or lead to the development of in vivo reprogramming strategies to regenerate the human heart.

6.2.2 microRNAs in Regulation of Pluripotency

The importance of miRNAs in the regulation of self-renewal and pluripotency came to light in studies with Dicer- and DGCR8-deficient mouse ESCs (mESCs) (Wang et al. 2007; Murchison et al. 2005) that lacked canonical miRNA processing. Disruption of both *Dicer* alleles resulted in embryonic lethality, indicating the importance of miRNAs in development. However, it is possible to generate proliferating Dicer-deficient mES cell lines, which indicates that canonical miRNA processing is not required for the maintenance of self-renewal. Nevertheless, Dicer-deficient mESCs do show severe defects in cell cycle progression and differentiation. This might be expected, since almost 50 % of all present miRNA molecules in mESCs, are produced from four loci (miR-21, miR-17–92 cluster, miR-15b-16 cluster and miR-290 cluster) involved in cell cycle regulation and oncogenesis (Calabrese et al. 2007). Similar to *Dicer* disruption, deletion of both *DGCR8* alleles resulted in mouse embryonic lethality early in development, again indicating the importance of canonical miRNA processing in development. Furthermore, DGCR8-deficient mES cell lines also displayed defects in differentiation and cell cycle progression (Wang et al. 2007). In general, loss of miRNAs in mESCs compromises the exit of self-renewal and progression of the cell cycle.

Further studies have focussed on miRNA profiling of mESCs and human ESCs (hESCs), which have revealed specific miRNA expression profiles. Interestingly, only a limited number of miRNAs are transcribed in a pluripotent stem cell state, and these are immediately silenced once cells receive differentiation signals (Houbaviy et al. 2003). Furthermore, reprogramming factors Oct4, Sox2, Nanog and c-Myc, which are able to induce pluripotency in somatic cells, occupy promoters of these miRNA families, hereby regulating their expression (Marson et al. 2008). The most abundant and ESC-specific miRNAs are transcribed from two miRNA clusters: the miR-290 (human homologues miR-371–373) cluster and the miR-302–367 cluster.

Transcription of miR-302–367 is regulated by Oct4 and Sox2 in hESCs and mESCs (Marson et al. 2008; Card et al. 2008). This miRNA cluster encodes miR-302a, miR302b, miR302c, miR-302d, and miR-367 (Suh et al. 2004), and promotes pluripotency by targeting several pathways. Lipchina et al. (2011) identified a list of 146 validated miR-302–367 targets through complementary experimental and computational approaches (Lipchina et al. 2011). The miR-302–367 cluster members silence inhibitors of cell survival and G1-S transition of the cell cycle. Repression of these targets contributes to the unusual short G1 cell cycle of ESCs, necessary to maintain rapid self-renewal of pluripotent stem cells (Wang et al. 2008a). Furthermore, miR-302–367 targets genes that promote heterochromatin formation to ensure an open chromatin formation, which is characteristic for pluripotent cells. Other targets of miR-302–367 include genes involved in metabolic regulation to promote glycolytic metabolism. By downregulating genes involved in oxidative phosphorylation and mitochondrial biogenesis, miR-302–367 promotes a metabolic state characteristic for ESCs (Lipchina et al. 2012). Another set of target genes of miR-302–367 is involved in intracellular vesicle transport and endocytosis. Although the exact role of intracellular vesicles and endocytosis in maintenance of self-renewal and pluripotency is unclear, studies have indicated that these processes play a role in proliferation (Szczyrba et al. 2011) and signalling pathways involved in development (Weigert et al. 2004).

The miR-290 cluster comprises the most highly expressed miRNAs in mESCs (>70 %) and possesses similar or identical seed sequences as the miR-302–367 cluster (Houbaviy et al. 2003). The human genome contains homologues of the miR-290 cluster, namely the miR-371–373 family (Wang et al. 2013). Besides the miR-302–367 cluster, certain members of the miR-290 cluster (miR-291a, miR-294, and miR-295) also regulate the G1-S phase transition during the cell cycle by primarily targeting cyclin-Cdk inhibitors (Wang et al. 2008a). Furthermore, Oct4 physically occupies the promoter of the miR-290 cluster, regulating its expression (Marson et al. 2008). MiR-290 targets include epigenetic repressive DNA methyltransferases, such as retinoblastoma-like protein 2 (Rbl2) (Sinkkonen et al. 2008). By downregulating Rbl2, this miRNA cluster promotes expression of Oct4, which in return promotes expression of the miR-290 cluster. Other targets of miR-290 include the NF-κB subunit p65, which normally promotes differentiation of ESCs (Lüningschrör et al. 2012). According to a recent study, the cluster also promotes glycolytic ESC metabolism by targeting Mbd2, which normally represses glycolytic metabolism (Cao et al. 2015). Though the miR-290 cluster is not expressed in hESCs, the homologous miR-371–373 family is predicted to target the same pathways in hESCs since almost all members from both clusters share the same seed sequence.

Interestingly, the seed sequence (AAGUGCU) of members of miR-290 and miR-302–367 clusters is shared by other miRNAs highly expressed in ESCs, including members of the miR-17–92 cluster, miR-106a–363 cluster, and miR-106b–25 cluster (Fig. 6.2) (Li and He 2012). Conservation of seed sequences between miRNA clusters expressed in ESCs indicates common mechanistic roles and partial redundancy of miRNAs in maintenance of self-renewal and pluripotency. However, these clusters also contain members with different seed sequences that might attribute to distinct characteristics of pluripotent stem cells of different species (Table 6.1) (Nichols and Smith 2009).

6.2.3 *microRNAs in Alternate States of Pluripotency*

Pluripotent stem cells of different species exhibit different characteristics and cellular responses. The 'naïve' pluripotent stem cell state is only represented by mESCs derived from the ICM of murine blastocysts. This state is characterized

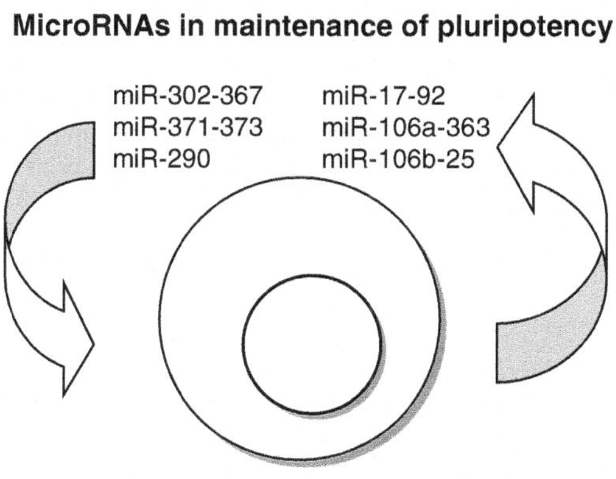

Fig. 6.2 MicroRNAs in the maintenance of pluripotency. The figure displays miRNA clusters that are involved in the maintenance of self-renewal and pluripotency. These miRNA clusters posses several members with similar or identical seed sequences (AAGUGCU), indicating common mechanistic roles of miRNAs in these processes

Table 6.1 MiRNA clusters expressed in mESCs and hESCs

miR-302-367 cluster (M, H)	
miR-302a	5′-UAAGUGCUUCCAUGUUUUGGUGA-3′
miR-302b	5′-UAAGUGCUUCCAUGUUUUAGUAG-3′
miR-302c	5′-AAGUGCUUCCAUGUUUCAGUGG-3′
miR-302d	5′-UAAGUGCUUCCAUGUUUGAGUGU-3′
miR-367	5′-AUUGCACUUUAGCAAUGGUGA-3′
miR-290 cluster (M)	
miR-290a	5′-ACUCAAACUAUGGGGGCACUUUU-3′
miR-291a-3p	5′-AAAGUGCUUCCACUUUGUGUGC-3′
miR-291b-3p	5′-AAAGUGCCGCCAGGUUUUGAGUGU-3′
miR-293-3p	5′-AGUGCCGCAGAGUUUGUAGUGU-3′
miR-294-3p	5′-AAAGUGCUUCCCUUUUGUGUGU-3′
miR-295-3p	5′-AAAGUGCUACUACUUUUGAGUCU-3′
miR-371-373 cluster (H)	
miR-371a-5p	5′-ACUCAAACUGUGGGGGCACUU-3′
miR-372-3p	5′-AAAGUGCUGCGACAUUUGAGCGU-3′
miR-373-3p	5′-GAAGUGCUUCGAUUUUGGGGUGU-3′
miR-17-92 cluster (M, H)	
miR-17-5p	5′-CAAAGUGCUUACAGUGCAGGUAG-3′
miR-20a-5p	5′-UAAAGUGCUUAUAGUGCAGGUAG-3′
miR-92a-3p	5′-UAUUGCACUUGUCCCGGCCU-3′
miR-106a-363 cluster (M, H)	
miR-20b-5p	5′-CAAAGUGCUCAUAGUGCAGGUAG-3′
miR-106a-5p	5′-CAAAGUGCUAACAGUGCAGGUAG-3′
miR-363-5p	5′-CAGGUGGAACACGAUGCAAUUUU-3′
miR-106b-25 cluster (M, H)	
miR-93-5p	5′-CAAAGUGCUGUUCGUGCAGGUAGU-3′
miR-106b-5p	5′-UAAAGUGCUGACAGUGCAGAUAG-3′
miR-25-5p	5′-AGGCGGAGACUUGGGCAAUUGCU-3′

MiR-290 and miR-302–367 clusters are highly expressed in mESCs, while miR-371–373 and miR-302–367 clusters are highly expressed in hESCs. These clusters share a common seed sequence (AAGUGCU), shown in red, which recognizes target mRNAs. Other clusters that share this seed sequence have also been identified, including miR-17–92, miR-106a-363, and miR-106b-25, and expression of these clusters is induced upon reprogramming with classic reprogramming factors [122]. In addition, these clusters, which are sometimes species-specific, each contain at least one member with a different seed sequence, suggesting a different function in pluripotency and self-renewal. If one pre-miRNA produces two mature miRNAs on its opposite arms in roughly similar amounts, addition of –3p or –5p suffix indicates which mature miRNA is referred to. *M*, mouse; *H*, human

by expression of Oct4 by a distal enhancer, global reduction of DNA methylation, and repressive chromatin marks on developmental regulatory gene promoters (Marks et al. 2012). In contrast, hESCs are termed 'primed' pluripotent cells, and display more similarity to cells derived from the post-implantation murine embryonic epiblast with regard to gene expression patterns, epigenetic state, and signalling response to maintain an undifferentiated state (Tesar et al. 2007).

These epiblast cells are called epiblast stem cells (EpiSCs) and are thought to represent a less pluripotent state than 'naïve' pluripotent cells, since they cannot contribute to blastocyst chimeras (Tesar et al. 2007). Properties of 'naïve' and 'primed' pluripotent stem cells are elaborately reviewed elsewhere (Nichols and Smith 2009).

A limited number of studies have investigated differences in miRNA expression profiles between alternate states of pluripotency. Nevertheless, these studies have demonstrated that mESCs predominantly express the miR-290 cluster, preceding miR-302–367 expression at the early epiblast stage later in development (Spruce et al. 2010). EpiSCs and hESCs predominantly express the miR-302–367 cluster (Kim et al. 2011). These findings indicate that miR-302–367 expression corresponds with a slightly less pluripotent cell state.

Several studies have suggested that human pluripotent stem cell lines (hESCs and human iPS cells) exhibit different developmental potential, which translates into varying differentiation propensity (Osafune et al. 2008), especially concerning neural cell types (Hu et al. 2010) or hemangioblastic lineages (Feng et al. 2010). Differentiation propensity is very important to consider when using cell sources for cardiac regeneration. In a study by Kim et al. (2011), hESCs and iPS cells that expressed high levels of the miR-371–373 cluster demonstrated higher level of 'naïve' ESC markers and greater neural differentiation propensity. Their results suggest that human pluripotent cell lines that highly express the miR-371–373 cluster may represent a more 'naïve' pluripotent state, corresponding with greater differentiation propensity (Kim et al. 2011).

Due to their high pluripotent state, self-renewal, and ability to differentiate into all cells of the cardiovascular lineage, ESCs and iPS cells appear to be the ideal cell source for cardiac regeneration. However, clinical use of pluripotent stem cells is hampered by significant obstacles, including safety concerns, immune rejection, and low efficiency of iPS cell generation (Nussbaum et al. 2007; Okano et al. 2013). Since miRNAs perform an important regulatory function in self-renewal and pluripotency, they may improve reprogramming efficiency or function as a strategy for iPS cell generation without the use of viral vectors.

6.2.4 microRNAs in Reprogramming

Transduction of fibroblasts with transcription factors OSKM directly affects expression of several miRNA clusters, including miR-290 (and human homolog miR-371–373) (Judson et al. 2009; Subramanyam et al. 2011), miR-302–367 (Subramanyam et al. 2011), and paralogous clusters miR-17–92, miR106b–25, and miR106a–363 (Li et al. 2011). Since c-Myc and Klf4 are involved in oncogenesis, they are best avoided during iPS cell generation when considering clinical applications (Dang 2012; Yu et al. 2011). Therefore, researchers have attempted to generate iPS cells and enhance reprogramming efficiency using aforementioned miRNAs in combination with reduced numbers of pluripotency factors.

Based on the high expression levels of the miR-290 cluster in mESCs, first attempts of miRNA-mediated reprogramming involved overexpression of different members of this cluster in mouse embryonic fibroblasts (Wang et al. 2008a). Addition of miR-291-3p, miR-294, and miR-295 increased the efficiency of reprogramming in combination with retroviral transfection of Oct4, Sox2, and Klf4 (OSK) (Judson et al. 2009). miR-294 demonstrated greatest increased efficiency of up to 75 % of that achieved with OSKM (Judson et al. 2009). Therefore, it was suggested that miR-294 can efficiently substitute for c-Myc when reprogramming fibroblasts into iPS cells.

Furthermore, members of the miR-302–367 cluster that share the same seed sequence with members of the miR-290 cluster, also demonstrated the ability to improve reprogramming efficiency (Judson et al. 2009). Recent studies suggest that miR-302 improves reprogramming efficiency by suppressing several targets, including a repressor of Oct3/4 (NR2F2) (Hu et al. 2013) and several epigenetic regulators (AOF2, AOF1, MECP1-p66, and MECP2) to promote Oct3/4 expression and open chromatin formation, respectively (Lin et al. 2011).

Three other miRNA clusters (miR-17–92 cluster, miR-106b–25 cluster, and the miR-106a–363 cluster), were shown to be highly expressed during early stages of reprogramming with OSKM (Li et al. 2011). Overexpression of miR-106b–25 members, miR-93 and miR-106b, increased reprogramming efficiency fourfold in fibroblasts transfected with OSK. Li et al. (2011) have shown that miR-93 and miR-106b target the TGFβ-receptor 2 (TGFbr2) and cyclin-Cdk inhibitor p21 (Li et al. 2011). TGFbr2 is a receptor kinase that induces epithelial-to-mesenchymal transition (EMT) upon TGF-β signalling, a process were epithelial cells lose their cell-cell contacts and gain migratory and invasive properties (Thiery et al. 2009). Fibroblasts, a product of EMT, can only be reverted to a pluripotent state by repressing pro-EMT signalling (i.e. TGF-β signalling), hereby initiating the reverse process, mesenchymal-to-epithelial transition (MET) (Li et al. 2010). Consistent to these findings, embryonic stem cells are morphologically similar to epithelial cells and express E-cadherin, an epithelial marker (Baum et al. 2008). Moreover, human miR-302b and human miR-372 were also found to inhibit TGFβ-signalling by targeting TGFbr2. Addition of these miRNAs to OSK or OSKM resulted in a 10- to 15-fold increase of reprogramming efficiency, respectively (Subramanyam et al. 2011).

In addition to miR-290, miR-302–367, miR-17–92, and miR-106b–25 cluster members, a recent library screen of 379 miRNAs observed improved reprogramming efficiency during reprogramming with OSK in combination with the miR-130–301–721 family (Pfaff et al. 2011). This newly identified miR-family is indirectly involved in cell cycle regulation by targeting transcription factor Meox2, which normally promotes expression of cell cycle inhibitor p21 (Pfaff et al. 2011).

Instead of resembling miRNA expression profiles of pluripotent cells, an alternative approach to promote reprogramming with miRNAs might be to reduce expression of miRNAs that are highly expressed in differentiated fibroblasts. Let-7 family miRNAs, which are specifically depleted in ESCs, repress pluripotency factors and are highly expressed during differentiation. Interestingly, let-7 and miR-290 cluster miRNAs perform opposite functions in ESC self-renewal and pluripotency (Melton et al. 2010). It has been demonstrated that Let-7 members directly target several ESC-specific genes, including c-Myc, and downregulate important cell cycle molecules such as Cdk4 and CyclinD, to repress the G1-S transition of the cell cycle (Melton et al. 2010; Schultz et al. 2008). Accordingly, let-7 miRNA inhibition resulted in a fourfold increase of reprogramming efficiency compared to three pluripotency factors (OSK) alone (Melton et al. 2010). Repression of tumor suppressor pathways also appears important during the generation of pluripotent cells. For instance, depletion of miR-34 (Choi et al. 2011) and inhibition of miR-199a-3p (Wang et al. 2012), which act downstream of tumor suppressor p53, enhanced OSKM reprogramming efficiency. Moreover, by directly targeting p53 mRNA, ectopic expression of mir-138 (Ye et al. 2012) also significantly improved the efficiency of iPS cell generation. However, p53 is shown to preserve genomic integrity by aborting reprogramming in cells carrying various types of DNA damage, normally resulting in p53-dependent apoptosis (Marión et al. 2009). This reserves a crucial role for p53 in preventing iPS cell generation from suboptimal cells (Marión et al. 2009). Therefore, p53-related miRNAs have to be further investigated before they can be safely used to enhance reprogramming. To date, repression of miRNAs specific for differentiated cells has only slightly increased the efficiency of reprogramming without transcription factor c-Myc (Yang et al. 2011).

Besides stimulating or repressing miRNAs to promote reprogramming efficiency, miRNAs have also been investigated to induce reprogramming without the addition of transcription factors. Only a few studies have reported miRNA-mediated reprogramming of somatic cells to iPS cells without additional reprogramming factors. Reprogramming in the absence of transcription factors was first achieved by expression of the miR-302–367 cluster in cancer cells (Lin et al. 2008). Further investigations of Lin *et al.* led to the generation of iPS cells through inducible miR-302s (a viral vector containing miR-302a, b, c and d) expression (Lin et al. 2011). Recently, Anokye-Danso et al. (2011) reported that overexpression of miR-302–367 cluster members successfully reprogrammed mouse and human fibroblasts into iPS cells in the absence of additional factors (Anokye-Danso et al. 2011). Direct transfection of mature double-stranded miR-200c in combination with miR-302 and miR-369 was also reported to be sufficient to induce pluripotency in somatic cells (Miyoshi et al. 2011). Lüningschrör et al. (2013) failed to reproduce these miRNA-mediated reprogramming experiments and could not induce pluripotency with miRNAs alone (Lüningschrör et al. 2013).

In summary, it has been demonstrated that miRNAs that target cell cycle inhibitors, epigenetic regulators, and modulators of mesenchymal-to-epithelial transition, clearly enhance reprogramming efficiency of somatic cells into iPS cells. Additional investigations are necessary to reveal if miRNAs alone can be used to effectively reprogram fibroblasts to iPS cells without the use of reprogramming factors (Fig. 6.3).

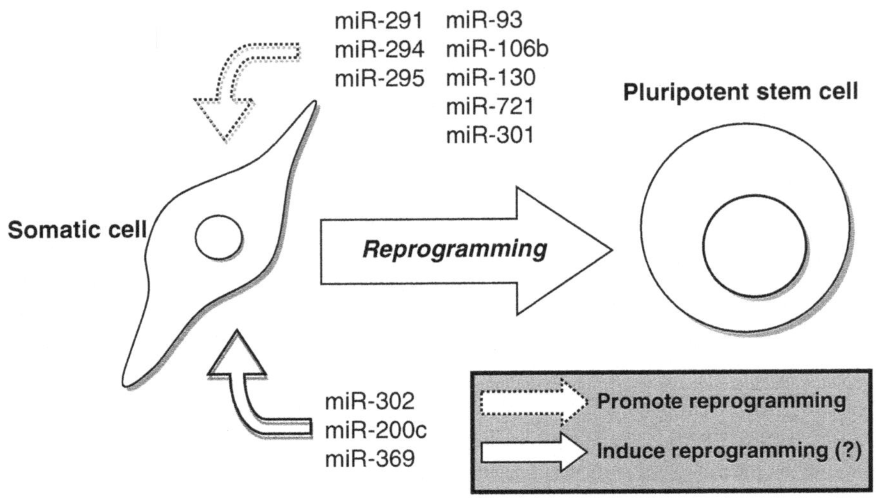

Fig. 6.3 MicroRNAs involved in reprogramming. The figure displays miRNAs that have demonstrated to promote reprogramming efficiency of somatic cells into iPS cells or were reported to induce reprogramming into iPS cells without additional transcription factors. However, the ability of miRNAs to reprogram cells without additional factors remains questionable

6.3 The Role of microRNAs in Cardiovascular Lineage Commitment

6.3.1 Introduction

Considering safety concerns with regard to pluripotent stem cells, differentiation of iPS cells and ESCs into specific lineages before transplantation is necessary. Canonical miRNAs play an essential role in cardiac development, as studies have shown that cardiac-specific knockout of the miRNA-processing enzyme Dicer results in rapid dilated cardiomyopathy, heart failure, and embryonic lethality (Chen et al. 2008). Since miRNAs are powerful regulators of gene expression patterns, they might represent an efficient strategy to modulate commitment of cardiac progenitor cells (CPCs) and pluripotent stem cells to cardiovascular lineages ex vivo or in vivo (summarised in Table 6.2).

6.3.2 microRNAs in Cardiomyocyte Differentiation and Proliferation

Specific miRNAs involved in cardiac differentiation have been revealed by miRNA expression profiling of hESC-derived cardiomyocytes and during differentiation of CPCs, which highlighted changes in several miRNAs. Especially expression of miR-1,

Table 6.2 MiRNAs in cardiovascular lineage commitment

Endothelial cells		Cardiomyocytes		Vascular smooth muscle cells	
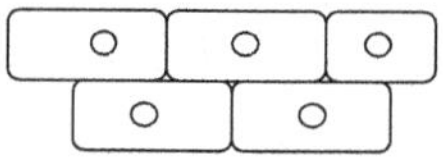		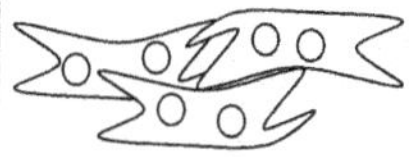		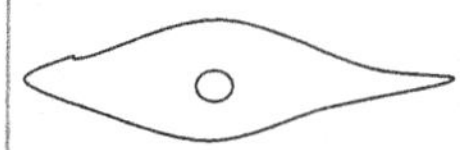	
miR-126	Essential for vascular integrity and angiogenesis	miR-1	Modulates proliferation and promotes cardiomyocyte differentiation	miR-1	Required for VSMC differentiation from ESCs
	Not sufficient to induce EC differentiation	miR-133	Represses smooth muscle genes	miR-10a	Inhibition blocks VSMC differentiation from ESCs
miR-17–92	Inhibition of miR-92a enhances neo-vascularisation		Overexpression inhibits cardiomyocyte differentiation		
	Blocks ischemia-induced angiogenesis, enhances tumour-associated angiogenesis	miR-208a	Regulates cardiomyocyte hypertrophy	miR-143/145	Promotes VSMC differentiation
miR-99b, miR-181a/b	Promote EC differentiation from ESCs	miR-499	Increases cardiomyocyte differentiation		Necessary to acquire and maintain contractile phenotype (maturation)
miR-130, miR-210 etc.	Several miRNAs promote angiogenesis and are upregulated during EC differentiation			miR-221/222	Promote VSMC proliferation by targeting cyclin-cdk inhibitors

Some miRNAs have been demonstrated to directly control differentiation of ESCs and/or cardiac progenitor cells. Other miRNAs play crucial roles in development and normal cellular functions

miR-133, miR-208, and miR-499, changed remarkably during cardiac differentiation (Sluijter et al. 2010; Ivey et al. 2008; Wilson et al. 2010).

miR-1 and miR-133 are highly expressed in cardiac and skeletal muscle cells, where they modulate muscle proliferation and differentiation while repressing other lineages (Ivey et al. 2008; Chen et al. 2006). In humans, miR-1 and miR-133 are under regulation of cardiac and muscle specific transcription factors, including serum response factor (SRF), myocyte enhancer factor 2 (MEF2), and myoblast determination protein (MyoD) (Liu et al. 2007; Zhao et al. 2005). Although miR-1 and miR-133 are transcribed together as a polycystronic cluster, they appear to perform distinct functions during proliferation and differentiation of myoblasts

(Chen et al. 2006). miR-1 promotes myogenesis by targeting histone deacetylase 4 (HDAC4), a transcriptional repressor of myogenic gene expression. Furthermore, by directly targeting frizzled family receptor 7 (FZD7) and fibroblast growth factor receptor substrate 2 (FRS2), miR-1 suppresses WNT and FGF signalling, which promotes cardiomyocyte differentiation (Lu et al. 2013). By targeting Hand2, a transcription factor required for ventricular cardiomyocyte proliferation, miR-1 also modulates the growth of the developing heart. Accordingly, miR-1 deficient mice displayed enhanced Hand2 expression, which resulted in an abnormally enlarged heart (Zhao et al. 2007). Consistently, miR-1 overexpression resulted in thin-walled ventricles of the heart in a mouse model, indicating premature cardiomyocytes due to early exit of the cell cycle and hampered cardiomyocyte proliferation (Zhao et al. 2005). Furthermore, miR-1 overexpression promotes cardiomyocyte differentiation from cardiac progenitor cells and ESCs of both mouse and human origin, demonstrated by an increase in cardiac marker expression (Sluijter et al. 2010; Lu et al. 2013).

In contrast to miR-1, which promotes differentiation, overexpression of miR-133 represses cardiac differentiation of hESCs and mESCs (Ivey et al. 2008). Furthermore, miR-133 directly targets muscle transcription factor SRF, suppressing smooth muscle gene expression (Liu et al. 2008). Accordingly, complete deletion of both mir-133 alleles results in ectopic activation of smooth muscle genes in the developing heart, as well as excessive proliferation and apoptosis of cardiomyocytes (Liu et al. 2008). Moreover, miR-133 overexpression in mouse models led to ventricular wall thinning and decreased cardiomyocyte proliferation (Liu et al. 2008; Carè et al. 2007). Altogether, studies have shown that polycystronic transcription of miR-1/miR-133 results in adequate regulation of cardiac proliferation and cardiomyocyte differentiation during cardiac development.

Further studies on miRNAs in cardiac development identified a family of intronic miRNAs, termed 'myomiRs', consisting of miR-499, miR-208a and miR-208b, which share many target genes. Although miR-208b and miR-499 are expressed in skeletal and cardiac muscle, miR-208a is only expressed in the heart where it regulates cardiomyocyte hypertrophy (Liu and Olson 2010; Callis et al. 2009). miR-499 is highly expressed in cardiac progenitor cells (Sluijter et al. 2010) and the heart (van Rooij et al. 2009), and in hESCs during cardiac differentiation (Wilson et al. 2010). Overexpression of miR-499 in human cardiac progenitor cells reduced proliferation of these cells and induced differentiation into cardiomyocytes (Sluijter et al. 2010). Furthermore, injection of miR-499 into the border zone of infracted rat hearts enhanced cardiomyogenesis in vivo (Sluijter et al. 2010). Consistent with these findings, expression of miR-499 resulted in upregulation of cardiac markers, such as Nkx2.5 and GATA4, by repression of SOX6 and PTBP3. In contrast, inhibition of miR-499 in mouse and human ESCs inhibits cardiomyocyte differentiation (Sluijter et al. 2010). In the mammalian heart, miR-499 is expressed at very low levels in c-kit positive cardiac stem cells, while it is highly expressed in cardiomyocytes, together with miR-1 and mir-133 (Hosoda et al. 2011). This suggests that these miRNAs are important in the specialization of cells towards a cardiomyocyte cell fate. Interestingly, miR-499 was demonstrated to translocate through gap junctions from cardiomyocytes to c-kit positive cardiac stem cells (CSCs) to promote differentiation into de novo cardiomyocytes (Hosoda 2013).

An alternative strategy to regenerate the heart is to stimulate endogenous cardiomyocyte proliferation. Several studies have shown that zebrafish are able to generate new myocardium from existing cardiomyocytes (Kikuchi et al. 2010; Jopling et al. 2010). During zebrafish heart regeneration, cardiomyocytes around the cardiac injury dedifferentiate into a more immature phenotype and cardiac transcription factors, such as Gata4 and Hand2, are activated (Kikuchi et al. 2010; Schindler et al. 2014). Subsequently, these dedifferentiated cardiomyocytes re-enter the cell cycle and fully repair the injured heart (Jopling et al. 2010). Decreased expression of miR-133 was observed by Yin et al. (2012) during zebrafish heart generation. Accordingly, miR-133 depletion enhanced cardiomyocyte proliferation by elevating repression of cell cycle factors and cell junction components (Yin et al. 2012). More recently, Aguirre et al. (2014) identified two miRNA families, miR-99/100 and let-7a/c, which were also downregulated during zebrafish heart regeneration. Alas, miR-99/100 expression remained unchanged after cardiac injury in mice. Nevertheless, knockdown of miR-99/100 and let-7a/c in mice reduced scar size and improved cardiac function after myocardial injury (Aguirre et al. 2014). Recently, direct injection of viral vectors expressing human miR-590 and miR-199a promoted cardiomyocyte proliferation and restored cardiac function in mice by targeting genes involved in cell cycle and proliferation (Eulalio et al. 2012). Altogether, these results indicate that regenerative mechanisms of the heart may be conserved between species, although they are not sufficiently stimulated upon cardiac injury in mammals.

6.3.3 microRNAs in Vascular Smooth Muscle Differentiation

The importance of miRNAs in vascular smooth muscle cell (VSMC) development was revealed during experiments with mutant mice that possessed a dysfunctional Dicer allele under a VSMC-specific promoter. Mutant mice died at embryonic day 16 or 17 due to bleeding and impaired VSMC contractility. Interestingly, these vascular defects observed in mutant mice could be partially rescued by overexpression of miR-145, highlighting the importance of this specific miRNA in VSMC development (Albinsson et al. 2010).

miR-145 is the most highly expressed miRNA in VSMCs, and is transcribed together with miR-143. During development, miR-145/143 is expressed in VSMCs under control of Nkx2.5, SRF, and its co-activator, myocardin. These miRNAs target a number of transcription factors, including Klf4 and Elk1 (both involved cellular proliferation), as well as a number of cytoskeletal proteins, to promote differentiation and repress proliferation of VSMCs (Cordes et al. 2009). In addition, miR-145, but not miR-143, is necessary for myocardin-induced reprogramming of fibroblasts into VSMCs and overexpression of miR-145 induced VSMC differentiation from neural crest cells (Cordes et al. 2009). Inhibition of miR-145 did not fully inhibit VSMC differentiation from ESCs but resulted in a more immature SMC phenotype (Cordes et al. 2009). Furthermore, miR-143/145 expression appears necessary for VSMCs to acquire and maintain a contractile phenotype (Boettger et al. 2009). Loss of these miRNAs results in impeded neointima formation because miR-143/145 act downstream of SRF during cytoskeletal remodelling and VSMC migration after vascular injury (Xin et al. 2009). Surprisingly, overexpression of miR-143 and miR-145 also decreased neointima formation in a rat model of acute vascular injury (Elia et al. 2009). These data indicate that miR-143 and miR-145 play crucial roles in modulating VSMC maturation and migration, though they are not essential for VSMC differentiation from stem cells.

Further studies have identified miR-1, which also mediates cardiomyocyte differentiation, as one of the regulators in VSMC differentiation (Xie et al. 2011). miR-1 is required for VSMC differentiation from ESCs, as inhibition of miR-1 led to a down-regulation of VSMC-specific markers and decreased the number of VSMCs from ESCs. Similar to miR-145, miR-1 also targets transcription factor Klf4 (Xie et al. 2011). In addition to miR-1, miRNA profiling during VSMC differentiation demonstrated an increased expression of miR-10a, which was shown to target HDAC4 (Huang et al. 2010). Accordingly, inhibition of miR-10a blocked VSMC differentiation from ESCs. However, it has not been revealed how repression of HDAC4 by miR-10a is specifically involved in VSMC differentiation. Furthermore, miR-221 and miR-222 were found to be involved in VSMC proliferation by targeting cyclin-dependent kinase inhibitors, p57 and p27 (Davis et al. 2009; Liu et al. 2009). Consistent with these findings, knockdown of miR-221 and miR-222 suppressed VSMC proliferation in vivo (Liu et al. 2009). More recent studies on VSMC differentiation and proliferation have identified other miRNAs that either promote or inhibit VSMC differentiation. miRNAs that promote differentiation include miR-663 (Li et al. 2013) and miR-18a-5p (Kee et al. 2014), while others such as miR-132 (Choe et al. 2013), and miR-365 (Zhang et al. 2014) inhibit VSMC differentiation. Further studies are necessary to discover the potential of these miRNAs with regard to heart regeneration after injury.

6.3.4 microRNAs in Endothelial Differentiation and Vascular Development

Early studies demonstrated that Dicer-deficient mouse embryos exhibited defects in vasculogenesis and angiogenesis, which was consistent with altered expression of vascular marker genes such as vascular endothelial growth factor (VEGF), vascular endothelial growth factor receptor-1 and receptor-2, and Tie1 (Yang et al. 2005). Later in development, these Dicer-deficient mice had an impaired angiogenic potential after vascular injury (Suárez et al. 2008).

The most extensively studied angiogenic miRNA is miR-126, which was demonstrated to be highly important for vessel integrity and angiogenic signalling by early studies in zebrafish and mice (Fish et al. 2008; Wang et al. 2008b). Targeted deletion of miR-126 resulted in a reduced angiogenic response to endothelial growth factors, loss of vascular integrity, and induced haemorrhages in zebrafish (Fish et al. 2008). In mice, loss of miR-126 also led to leaky blood vessels, emphasising the importance of miR-126 in vascular physiology. As identified by microarray analysis, loss of angiogenic potential and vascular integrity can, at least in part, be attributed to increased levels of Spred1 and Pik3r2, which are repressors of angiogenic signalling and direct targets of miR-126 (Fish et al. 2008; Wang et al. 2008b). Moreover, miR-126 has been demonstrated to modulate expression of stromal cell-derived factor 1 (SDF-1) in ECs, which is proposed to regulate recruitment of vascular progenitor cells to ischemic regions (van Solingen et al. 2011). In addition to pro-angiogenic effects, miR-126 regulates expression of inflammatory mediators in endothelial cells, such as vascular cell adhesion molecule 1 (VCAM-1), hereby limiting leukocyte adhesion and subsequent inflammation (Harris et al. 2008). Although enriched in endothelial progenitor cells (EPCs), miR-126 lacks the potential to induce endothelial cell differentiation in ESCs, as expression of several vascular marker genes was not affected by miR-126 overexpression (Fish et al. 2008). In summary, miR-126 is important for vessel integrity and angiogenesis by modulating angiogenic signalling, vascular progenitor cell recruitment, and inflammation. However, miR-126 is not sufficient to induce EC differentiation in ESCs, indicating a minor role in endothelial lineage commitment.

The miR-17–92 cluster, induced by transcription factor c-Myc, comprises of miR-17, -18a, -19a/b, -20a, and -92a, and is also expressed in ECs. Interestingly, this cluster has shown to perform distinct functions in different contexts. First, the miR-17–92 cluster plays an important role in angiogenesis after vascular injury (O'Donnell et al. 2005; He et al. 2005). One of the clusters members, miR-92a, performs anti-angiogenic functions in ECs by targeting several proangiogenic factors. Accordingly, inhibition of miR-92a enhanced neovascularisation and functional recovery after myocardial infarction (Bonauer et al. 2009). Furthermore, expression of miR-17, miR-18a, miR-19a, and miR-20a, has shown to inhibit angiogenic activity of ECs in vitro, while inhibition of these miRNAs led to stimulation of angiogenic processes (Doebele et al. 2010). In contrast, miR-17–92 cluster appears to enhance neovascularisation in tumour cells. Therefore, it is proposed that the effect of miR-17–92 cluster differs for ischemia-induced angiogenesis and tumour-associated angiogenesis, which is important to consider when developing clinical therapies (Dews et al. 2006).

miRNA profiling during EC differentiation from ESCs further revealed several miRNAs essential for this process (Kane et al. 2010, 2012; Luo et al. 2013). Some miRNAs that significantly change in their expression during EC differentiation either promote angiogenesis (let7b, let7f, miR-126, miR-130a, mir-133/b, miR-210, and miR-296), or impair angiogenesis (miR-20a/b, miR-221, and miR-222), and are reviewed elsewhere (Wu et al. 2009). To date, not all of these miRNAs have been tested for their potential to induce EC differentiation.

Identified as one of the first mammalian miRNAs, miR-21 was found to induce stem cell differentiation by modulation of TGF-β2 signalling (Di Bernardini et al. 2014). In a recent study on EC differentiation from iPS cells, miR-21 overexpression in iPS cells exposed to VEGF induced capillary formation in vitro and in vivo (Di Bernardini et al. 2014). Recently, miRNA gain/loss-of-function analysis identified two other miRNAs involved in EC differentiation from hESCs (Luo et al. 2013). These miRNAs, miR-150 and miR-200c, are proposed to be involved in EC differentiation by repressing transcriptional repressor zinc finger E-box-binding homeobox 1 (ZEB1), which has been identified as a repressor for EC gene expression (Luo et al. 2013). Furthermore, Kane et al. (2012) indicated that miR-99b, -181a, and -181b are also involved in EC differentiation (Kane et al. 2012). Expression levels of these miRNAs increased in a time- and differentiation-dependent manner and peaked in hESC-derived ECs and adult ECs. Lentiviral-mediated transfer of these miRNAs promoted EC-specific markers, as well as neovascularisation after transplantation with hESC-derived ECs in vivo (Kane et al. 2012). However, knockdown of miR-99b, -181a, and -181b did not inhibit EC differentiation, suggesting that these miRNAs are not indispensible in endothelial commitment.

Although several miRNAs in EC differentiation have been described here, many other miRNAs perform regulatory functions in vascular development, angiogenesis and endothelial function, including miR-221 (Nicoli et al. 2012), miR-132 (Anand et al. 2010), miR-218 (Small et al. 2010), miR-23–27–24 cluster (Zhou et al. 2011), and miR-27a/b (Urbich et al. 2012). The role of these miRNAs in vascular development is extensively reviewed elsewhere (Dang et al. 2013).

6.4 The Role of microRNAs in Direct Reprogramming

As mentioned before, differentiation of pluripotent stem cells before transplantation could represent a safe strategy for cardiac regeneration. However, differentiation protocols often lead to cardiac cells with varying phenotypes, which hampers mechanical and electrical coupling of transplanted cells, especially cardiomyocytes, with the surrounding tissue (Ng et al. 2010). This has forced researchers to develop new strategies for cardiac regeneration, including stimulating cardiac cell populations in situ to promote regeneration. A groundbreaking discovery was made by Ieda et al., who directly converted mouse cardiac fibroblasts into induced cardiomyocyte-like cells (iCMs) with a combination of three cardiac developmental transcription factors, Gata4, Mef2c, and Tbx5 (GMT) (Ieda et al. 2010). By avoiding a pluripotent cell state, some limitations of undifferentiated iPS cells, such as tumorigenicity, can be overcome. Reprogrammed cells expressed cardiac-specific markers, cardiomyocyte gene expression profiles, and exhibited spontaneous contraction. Further studies also used alternative combinations of transcription factors to reprogram cardiac fibroblasts into cardiomyocytes (Song et al. 2012; Protze et al. 2012). In addition, reprogramming efficiency into functional cardiomyocytes could be improved by addition of miR-133, which generated sevenfold more beating mouse cardiomyocyte-like cells than reprogramming with transcription factors alone (Muraoka et al. 2014).

Recently, others successfully achieved microRNA-mediated reprogramming of cardiac fibroblasts into cardiomyocytes via lentiviral transfection with cardiac-specific miRNAs without the use of additional transcription factors (Jayawardena et al. 2012). Four miRNAs, miR-1, miR-133, miR-208, and miR-499, highly expressed in cardiomyocytes, and tightly involved in cardiac development (Porrello 2013), were used by Jayawardena et al. (2012) to convert mouse cardiac fibroblasts into cardiomyocytes in vitro and in vivo (Jayawardena et al. 2012). Reprogramming

Table 6.3 Direct reprogramming in vivo and in vitro into induced cardiomyocyte-like cells (iCMs)

Reprogramming factors	Organism	In vitro/In vivo
Gata4, Mef2c, Tbx5	Mouse	In vitro: 0.5 % beating iCMs
		In vivo: >50 % beating iCMs
Gata4, Mef2c, Tbx5, Hand2	Mouse	In vitro: 0.2 % beating iCMs
		In vivo: beating iCMs
Gata4, Mef2c, Tbx5, Thymosin β4	Mouse	In vivo: beating iCMs
Gata4, Mef2c, Tbx5, *mir-133*	Mouse	In vitro: rare beating iCMs
Gata4, Hand2, Myocardin, Tbx5, *miR-1*, *miR-133*	Human	In vitro: rare beating iCMs
miR-1, *miR-133*, *miR-208*, *miR-499*	Mouse	In vitro: 1 % beating iCMs
		In vivo: beating iCMs

Different approaches with transcription factors and/or miRNAs have been used to induce iCM differentiation from cardiac fibroblasts for cardiac regeneration

of cardiac fibroblasts into cardiomyocytes resulted in more beating iCMs in vivo than in vitro, suggesting that other unknown factors in the heart enhance maturation of the iCMs. Reprogramming efficiency of cardiac fibroblasts into cardiomyocytes remains low (12 %) in this initial phase of miRNA-mediated direct reprogramming (Jayawardena et al. 2015). However, recent studies have shown that miRNAs are a promising alternative for direct reprogramming of somatic cells (Table 6.3).

6.5 Future Perspectives

Current understanding suggests that the mammalian heart possesses an endogenous regenerative capacity that could be stimulated to regenerate the heart after injury. The ideal cell source for cell-based cardiac regeneration should be able to differentiate into cardiomyocytes that integrate with existing cardiomyocytes inside the heart and differentiate into new blood vessels to supply blood to the infarct zone. Although several stem cell sources have shown little cardiac differentiation potential and predominantly promote cardiac function through paracrine signalling, pluripotent stem cells and cardiac progenitor cells are capable to differentiate into cardiovascular lineages in vitro and in vivo. Despite the enormous potential of autologous iPS cells in personal regenerative medicine, significant safety concerns remain with respect to the generation and transplantation of iPS cells.

In this chapter, microRNAs have been described as powerful regulators of gene expression patterns, regulating biological processes and cellular behaviour, including stem cell pluripotency, self-renewal, reprogramming, and cardiovascular lineage commitment. Due to increasing knowledge of miRNA expression patterns in different cell types, researchers have identified specific miRNAs that are able to promote reprogramming and induce differentiation of pluripotent stem cells and cardiac progenitor cells ex vivo. Because miRNAs are small, easily synthesized,

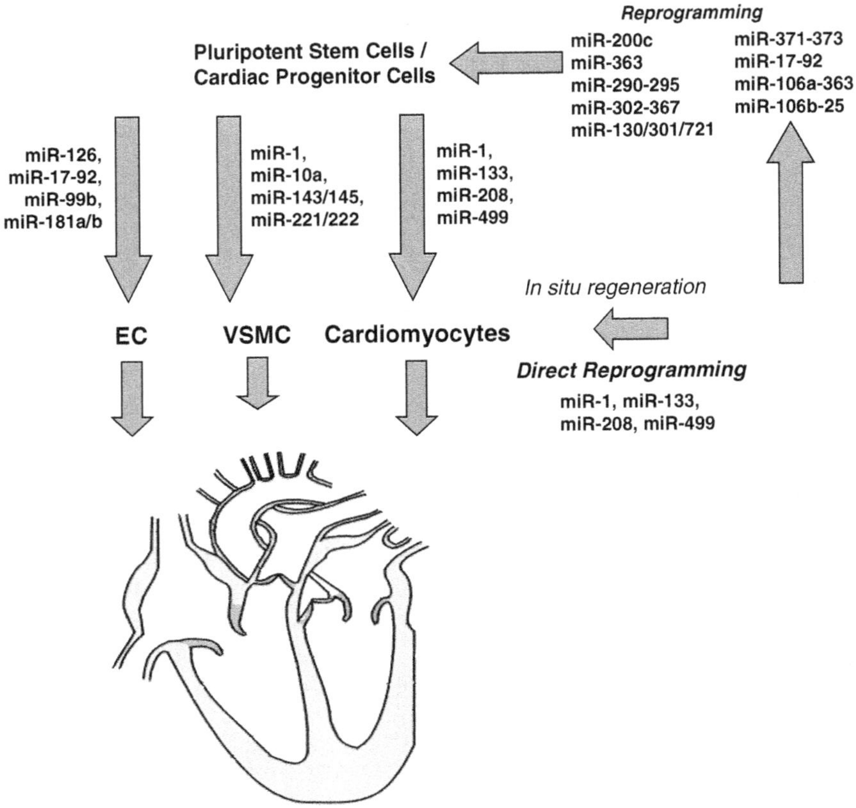

Fig. 6.4 MiRNA-mediated cardiac regeneration. MiRNAs are involved in maintenance of self-renewal and pluripotency, as well as reprogramming and differentiation. miRNAs control gene expression patterns, enabling the use of miRNAs to direct cellular behaviour. Reprogramming efficiency of fibroblasts into iPS cells can be increased by several miRNAs. Subsequently, pluripotent stem cells can be efficiently differentiated into cardiovascular cell types that can be transplanted to replenish lost cells after myocardial infarction. Furthermore, differentiation of resident cardiac progenitor cells into cardiovascular lineages might also be stimulated by miRNAs to promote cardiac regeneration. Recent studies demonstrated that miRNAs are also sufficient to induce direct reprogramming of cardiac fibroblasts into functional cardiomyocytes

and administered to cells by lipid-based transfection, they represent useful tools for the development of safe cell-based therapies to regenerate the heart.

Recent studies have reported the possibility to reprogram cardiac fibroblasts directly into beating cardiomyocytes ex vivo and in vivo. Interestingly, the efficiency of this process can be increased by miRNAs and a combination of miR-1, miR-133, miR-208, and miR-499 induced reprogramming of cardiac fibroblasts into functional cardiomyocytes without additional factors. This indicates that miRNAs are able to generate new cardiomyocytes from fibroblasts, hereby promoting in situ regeneration (Fig. 6.4). However, before miRNAs can be used in clinical settings for cardiac regeneration, safe and efficient delivery methods have to be developed (Zhang et al. 2013). By further understanding the role of specific miRNAs in reprogramming, and pluripotent stem cell or cardiac progenitor cell differentiation, clinical therapies could be developed to stimulate endogenous repair mechanisms and in situ regeneration of the human heart.

Acknowledgements This research forms part of the Project P1.05 LUST of the research program of the BioMedical Materials institute, co-funded by the Dutch Ministry of Economic Affairs and the Netherlands CardioVascular Research Initiative (CVON): the Dutch Heart Foundation, Dutch Federation of University Medical Centers, the Netherlands Organization for Health Research and Development, and the Royal Netherlands Academy of Sciences.

References

Abrahante JE, Daul AL, Li M, Volk ML, Tennessen JM, Miller EA, Rougvie AE (2003) The Caenorhabditis elegans hunchback-like gene lin-57/hbl-1 controls developmental time and is regulated by microRNAs. Dev Cell 4(5):625–637

Aguirre A, Montserrat N, Zachiggna S, Nivet E, Hishida T, Krause MN, Kurian L, Ocampo A, Vazquez-Ferrer E, Rodriguez-Esteban C, Kumar S, Moresco JJ, Yates JR, Campistol JM, Sancho-Martinez I, Giacca M, Izpisua Belmonte JC (2014) In vivo activation of a conserved microRNA program induces mammalian heart regeneration. Cell Stem Cell 15(5):589–604

Albinsson S, Suarez Y, Skoura A, Offermanns S, Miano JM, Sessa WC (2010) MicroRNAs are necessary for vascular smooth muscle growth, differentiation, and function. Arterioscler Thromb Vasc Biol 30(6):1118–1126

Anand S, Majeti BK, Acevedo LM, Murphy EA, Mukthavaram R, Scheppke L, Huang M, Shields DJ, Lindquist JN, Lapinski PE, King PD, Weis SM, Cheresh DA (2010) MicroRNA-132-mediated loss of p120RasGAP activates the endothelium to facilitate pathological angiogenesis. Nat Med 16(8):909–914

Anokye-Danso F, Trivedi CM, Juhr D, Gupta M, Cui Z, Tian Y, Zhang Y, Yang W, Gruber PJ, Epstein JA, Morrisey EE (2011) Highly efficient miRNA-mediated reprogramming of mouse and human somatic cells to pluripotency. Cell Stem Cell 8(4):376–388

Balsam LB, Wagers AJ, Christensen JL, Kofidis T, Weissman IL, Robbins RC (2004) Haematopoietic stem cells adopt mature haematopoietic fates in ischaemic myocardium. Nature 428(6983):668–673

Baum B, Settleman J, Quinlan MP (2008) Transitions between epithelial and mesenchymal states in development and disease. Semin Cell Dev Biol 19(3):294–308

Béthune J, Artus-Revel CG, Filipowicz W (2012) Kinetic analysis reveals successive steps leading to miRNA-mediated silencing in mammalian cells. EMBO Rep 13(8):716–723

Bhattacharyya SN, Habermacher R, Martine U, Closs EI, Filipowicz W (2006) Relief of microRNA-mediated translational repression in human cells subjected to stress. Cell 125(6): 1111–1124

Blau HM, Chiu CP, Webster C (1983) Cytoplasmic activation of human nuclear genes in stable heterocaryons. Cell 32(4):1171–1180

Boettger T, Beetz N, Kostin S, Schneider J, Krüger M, Hein L, Braun T (2009) Acquisition of the contractile phenotype by murine arterial smooth muscle cells depends on the Mir143/145 gene cluster. J Clin Invest 119(9):2634–2647

Bonauer A, Carmona G, Iwasaki M, Mione M, Koyanagi M, Fischer A, Burchfield J, Fox H, Doebele C, Ohtani K, Chavakis E, Potente M, Tjwa M, Urbich C, Zeiher AM, Dimmeler S (2009) MicroRNA-92a controls angiogenesis and functional recovery of ischemic tissues in mice. Science 324(5935):1710–1713

Braun JE, Huntzinger E, Fauser M, Izaurralde E (2011) GW182 proteins directly recruit cytoplasmic deadenylase complexes to miRNA targets. Mol Cell 44(1):120–133

Braun JE, Truffault V, Boland A, Huntzinger E, Chang C-T, Haas G, Weichenrieder O, Coles M, Izaurralde E (2012) A direct interaction between DCP1 and XRN1 couples mRNA decapping to 5′ exonucleolytic degradation. Nat Struct Mol Biol 19(12):1324–1331

Cai X, Hagedorn CH, Cullen BR (2004) Human microRNAs are processed from capped, polyadenylated transcripts that can also function as mRNAs. RNA 10(12):1957–1966

Calabrese JM, Seila AC, Yeo GW, Sharp PA (2007) RNA sequence analysis defines Dicer's role in mouse embryonic stem cells. Proc Natl Acad Sci U S A 104(46):18097–18102

Callis TE, Pandya K, Seok HY, Tang R-H, Tatsuguchi M, Huang Z-P, Chen J-F, Deng Z, Gunn B, Shumate J, Willis MS, Selzman CH, Wang D-Z (2009) MicroRNA-208a is a regulator of cardiac hypertrophy and conduction in mice. J Clin Invest 119(9):2772–2786

Campbell KH, McWhir J, Ritchie WA, Wilmut I (1996) Sheep cloned by nuclear transfer from a cultured cell line. Nature 380(6569):64–66

Cao Y, Guo W-T, Tian S, He X, Wang X-W, Liu X, Gu K-L, Ma X, Huang D, Hu L, Cai Y, Zhang H, Wang Y, Gao P (2015) miR-290/371-Mbd2-Myc circuit regulates glycolytic metabolism to promote pluripotency. EMBO J 34(5):609–623

Card DAG, Hebbar PB, Li L, Trotter KW, Komatsu Y, Mishina Y, Archer TK (2008) Oct4/Sox2-regulated miR-302 targets cyclin D1 in human embryonic stem cells. Mol Cell Biol 28(20):6426–6438

Carè A, Catalucci D, Felicetti F, Bonci D, Addario A, Gallo P, Bang M-L, Segnalini P, Gu Y, Dalton ND, Elia L, Latronico MVG, Høydal M, Autore C, Russo MA, Dorn GW, Ellingsen O, Ruiz-Lozano P, Peterson KL, Croce CM, Peschle C, Condorelli G (2007) MicroRNA-133 controls cardiac hypertrophy. Nat Med 13(5):613–618

Chen J-F, Mandel EM, Thomson JM, Wu Q, Callis TE, Hammond SM, Conlon FL, Wang D-Z (2006) The role of microRNA-1 and microRNA-133 in skeletal muscle proliferation and differentiation. Nat Genet 38(2):228–233

Chen J-F, Murchison EP, Tang R, Callis TE, Tatsuguchi M, Deng Z, Rojas M, Hammond SM, Schneider MD, Selzman CH, Meissner G, Patterson C, Hannon GJ, Wang D-Z (2008) Targeted deletion of Dicer in the heart leads to dilated cardiomyopathy and heart failure. Proc Natl Acad Sci U S A 105(6):2111–2116

Chendrimada TP, Gregory RI, Kumaraswamy E, Norman J, Cooch N, Nishikura K, Shiekhattar R (2005) TRBP recruits the Dicer complex to Ago2 for microRNA processing and gene silencing. Nature 436(7051):740–744

Choe N, Kwon J-S, Kim J-R, Eom GH, Kim Y, Nam K-I, Ahn Y, Kee HJ, Kook H (2013) The microRNA miR-132 targets Lrrfip1 to block vascular smooth muscle cell proliferation and neointimal hyperplasia. Atherosclerosis 229(2):348–355

Choi YJ, Lin C-P, Ho JJ, He X, Okada N, Bu P, Zhong Y, Kim SY, Bennett MJ, Chen C, Ozturk A, Hicks GG, Hannon GJ, He L (2011) miR-34 miRNAs provide a barrier for somatic cell reprogramming. Nat Cell Biol 13(11):1353–1360

Cordes KR, Sheehy NT, White MP, Berry EC, Morton SU, Muth AN, Lee T-H, Miano JM, Ivey KN, Srivastava D (2009) miR-145 and miR-143 regulate smooth muscle cell fate and plasticity. Nature 460(7256):705–710

Dang CV (2012) MYC on the path to cancer. Cell 149(1):22–35

Dang LTH, Lawson ND, Fish JE (2013) MicroRNA control of vascular endothelial growth factor signaling output during vascular development. Arterioscler Thromb Vasc Biol 33(2):193–200

Davis BN, Hilyard AC, Nguyen PH, Lagna G, Hata A (2009) Induction of microRNA-221 by platelet-derived growth factor signaling is critical for modulation of vascular smooth muscle phenotype. J Biol Chem 284(6):3728–3738

Dews M, Homayouni A, Yu D, Murphy D, Sevignani C, Wentzel E, Furth EE, Lee WM, Enders GH, Mendell JT, Thomas-Tikhonenko A (2006) Augmentation of tumor angiogenesis by a Myc-activated microRNA cluster. Nat Genet 38(9):1060–1065

Di Bernardini E, Campagnolo P, Margariti A, Zampetaki A, Karamariti E, Hu Y, Xu Q (2014) Endothelial lineage differentiation from induced pluripotent stem cells is regulated by microRNA-21 and transforming growth factor β2 (TGF-β2) pathways. J Biol Chem 289(6):3383–3393

Ding XC, Grosshans H (2009) Repression of C. elegans microRNA targets at the initiation level of translation requires GW182 proteins. EMBO J 28(3):213–222

Djuranovic S, Nahvi A, Green R (2012) miRNA-mediated gene silencing by translational repression followed by mRNA deadenylation and decay. Science 336(6078):237–240

Doebele C, Bonauer A, Fischer A, Scholz A, Reiss Y, Urbich C, Hofmann W-K, Zeiher AM, Dimmeler S (2010) Members of the microRNA-17-92 cluster exhibit a cell-intrinsic antiangiogenic function in endothelial cells. Blood 115(23):4944–4950

Eichhorn SW, Guo H, McGeary SE, Rodriguez-Mias RA, Shin C, Baek D, Hsu S-H, Ghoshal K, Villén J, Bartel DP (2014) mRNA Destabilization Is the Dominant Effect of Mammalian MicroRNAs by the Time Substantial Repression Ensues. Mol Cell 56(1):104–115

Elia L, Quintavalle M, Zhang J, Contu R, Cossu L, Latronico MVG, Peterson KL, Indolfi C, Catalucci D, Chen J, Courtneidge SA, Condorelli G (2009) The knockout of miR-143 and -145 alters smooth muscle cell maintenance and vascular homeostasis in mice: correlates with human disease. Cell Death Differ 16(12):1590–1598

Eulalio A, Huntzinger E, Nishihara T, Rehwinkel J, Fauser M, Izaurralde E (2009) Deadenylation is a widespread effect of miRNA regulation. RNA 15(1):21–32

Eulalio A, Mano M, Dal Ferro M, Zentilin L, Sinagra G, Zacchigna S, Giacca M (2012) Functional screening identifies miRNAs inducing cardiac regeneration. Nature 492(7429):376–381

Evans MJ, Kaufman MH (1981) Establishment in culture of pluripotential cells from mouse embryos. Nature 292(5819):154–156

Feng Q, Lu S-J, Klimanskaya I, Gomes I, Kim D, Chung Y, Honig GR, Kim K-S, Lanza R (2010) Hemangioblastic derivatives from human induced pluripotent stem cells exhibit limited expansion and early senescence. Stem Cells 28(4):704–712

Fish JE, Santoro MM, Morton SU, Yu S, Yeh R-F, Wythe JD, Ivey KN, Bruneau BG, Stainier DYR, Srivastava D (2008) miR-126 regulates angiogenic signaling and vascular integrity. Dev Cell 15(2):272–284

Friedman RC, Farh KK-H, Burge CB, Bartel DP (2009) Most mammalian mRNAs are conserved targets of microRNAs. Genome Res 19(1):92–105

Fukao A, Mishima Y, Takizawa N, Oka S, Imataka H, Pelletier J, Sonenberg N, Thoma C, Fujiwara T (2014) MicroRNAs trigger dissociation of eIF4AI and eIF4AII from target mRNAs in humans. Mol Cell 56(1):79–89

Guo H, Ingolia NT, Weissman JS, Bartel DP (2010) Mammalian microRNAs predominantly act to decrease target mRNA levels. Nature 466(7308):835–840

Gurdon JB, Elsdale TR, Fischberg M (1958) Sexually mature individuals of Xenopus laevis from the transplantation of single somatic nuclei. Nature 182(4627):64–65

Han J, Lee Y, Yeom K-H, Kim Y-K, Jin H, Kim VN (2004) The Drosha-DGCR8 complex in primary microRNA processing. Genes Dev 18(24):3016–3027

Han J, Lee Y, Yeom K-H, Nam J-W, Heo I, Rhee J-K, Sohn SY, Cho Y, Zhang B-T, Kim VN (2006) Molecular basis for the recognition of primary microRNAs by the Drosha-DGCR8 complex. Cell 125(5):887–901

Harris TA, Yamakuchi M, Ferlito M, Mendell JT, Lowenstein CJ (2008) MicroRNA-126 regulates endothelial expression of vascular cell adhesion molecule 1. Proc Natl Acad Sci U S A 105(5):1516–1521

He L, Thomson JM, Hemann MT, Hernando-Monge E, Mu D, Goodson S, Powers S, Cordon-Cardo C, Lowe SW, Hannon GJ, Hammond SM (2005) A microRNA polycistron as a potential human oncogene. Nature 435(7043):828–833

Hendrickson DG, Hogan DJ, McCullough HL, Myers JW, Herschlag D, Ferrell JE, Brown PO (2009) Concordant regulation of translation and mRNA abundance for hundreds of targets of a human microRNA. PLoS Biol 7(11), e1000238

Hertel J, Bartschat S, Wintsche A, Otto C, Stadler PF (2012) Evolution of the let-7 microRNA family. RNA Biol 9(3):231–241
Hosoda T (2013) The mircrine mechanism controlling cardiac stem cell fate. Front Genet 4:204
Hosoda T, Zheng H, Cabral-da-Silva M, Sanada F, Ide-Iwata N, Ogorek B, Ferreira-Martins J, Arranto C, D'Amario D, del Monte F, Urbanek K, D'Alessandro DA, Michler RE, Anversa P, Rota M, Kajstura J, Leri A (2011) Human cardiac stem cell differentiation Is regulated by a mircrine mechanism. Circulation 123(12):1287–1296
Houbaviy HB, Murray MF, Sharp PA (2003) Embryonic stem cell-specific MicroRNAs. Dev Cell 5(2):351–358
Hu B-Y, Weick JP, Yu J, Ma L-X, Zhang X-Q, Thomson JA, Zhang S-C (2010) Neural differentiation of human induced pluripotent stem cells follows developmental principles but with variable potency. Proc Natl Acad Sci U S A 107(9):4335–4340
Hu S, Wilson KD, Ghosh Z, Han L, Wang Y, Lan F, Ransohoff KJ, Burridge P, Wu JC (2013) MicroRNA-302 Increases Reprogramming Efficiency via Repression of NR2F2. Stem Cells 31(2):259–268
Huang H, Xie C, Sun X, Ritchie RP, Zhang J, Chen YE (2010) miR-10a contributes to retinoid acid-induced smooth muscle cell differentiation. J Biol Chem 285(13):9383–9389
Huntzinger E, Izaurralde E (2011) Gene silencing by microRNAs: contributions of translational repression and mRNA decay. Nat Rev Genet 12(2):99–110
Ieda M, Fu J-D, Delgado-Olguin P, Vedantham V, Hayashi Y, Bruneau BG, Srivastava D (2010) Direct reprogramming of fibroblasts into functional cardiomyocytes by defined factors. Cell 142(3):375–386
Ivey KN, Muth A, Arnold J, King FW, Yeh R-F, Fish JE, Hsiao EC, Schwartz RJ, Conklin BR, Bernstein HS, Srivastava D (2008) MicroRNA regulation of cell lineages in mouse and human embryonic stem cells. Cell Stem Cell 2(3):219–229
Jayawardena TM, Egemnazarov B, Finch EA, Zhang L, Payne JA, Pandya K, Zhang Z, Rosenberg P, Mirotsou M, Dzau VJ (2012) MicroRNA-mediated in vitro and in vivo direct reprogramming of cardiac fibroblasts to cardiomyocytes. Circ Res 110(11):1465–1473
Jayawardena TM, Finch EA, Zhang L, Zhang H, Hodgkinson CP, Pratt RE, Rosenberg PB, Mirotsou M, Dzau VJ (2015) MicroRNA induced cardiac reprogramming in vivo: evidence for mature cardiac myocytes and improved cardiac function. Circ Res 116(3):418–424
Jopling C, Sleep E, Raya M, Martí M, Raya A, Izpisúa Belmonte JC (2010) Zebrafish heart regeneration occurs by cardiomyocyte dedifferentiation and proliferation. Nature 464(7288):606–609
Judson RL, Babiarz JE, Venere M, Blelloch R (2009) Embryonic stem cell-specific microRNAs promote induced pluripotency. Nat Biotechnol 27(5):459–461
Kane NM, Meloni M, Spencer HL, Craig MA, Strehl R, Milligan G, Houslay MD, Mountford JC, Emanueli C, Baker AH (2010) Derivation of endothelial cells from human embryonic stem cells by directed differentiation: analysis of microRNA and angiogenesis in vitro and in vivo. Arterioscler Thromb Vasc Biol 30(7):1389–1397
Kane NM, Howard L, Descamps B, Meloni M, McClure J, Lu R, McCahill A, Breen C, Mackenzie RM, Delles C, Mountford JC, Milligan G, Emanueli C, Baker AH (2012) Role of microRNAs 99b, 181a, and 181b in the differentiation of human embryonic stem cells to vascular endothelial cells. Stem Cells 30(4):643–654
Kee HJ, Kim GR, Cho S-N, Kwon J-S, Ahn Y, Kook H, Jeong MH (2014) miR-18a-5p MicroRNA Increases Vascular Smooth Muscle Cell Differentiation by Downregulating Syndecan4. Korean Circ J 44(4):255–263
Kikuchi K, Holdway JE, Werdich AA, Anderson RM, Fang Y, Egnaczyk GF, Evans T, Macrae CA, Stainier DYR, Poss KD (2010) Primary contribution to zebrafish heart regeneration by gata4(+) cardiomyocytes. Nature 464(7288):601–605
Kim H, Lee G, Ganat Y, Papapetrou EP, Lipchina I, Socci ND, Sadelain M, Studer L (2011) miR-371-3 expression predicts neural differentiation propensity in human pluripotent stem cells. Cell Stem Cell 8(6):695–706
Kozomara A, Griffiths-Jones S (2014) miRBase: annotating high confidence microRNAs using deep sequencing data. Nucleic Acids Res 42(Database issue):D68–D73

Lagos-Quintana M, Rauhut R, Lendeckel W, Tuschl T (2001) Identification of novel genes coding for small expressed RNAs. Science 294(5543):853–858
Lee RC, Ambros V (2001) An extensive class of small RNAs in Caenorhabditis elegans. Science 294(5543):862–864
Lee RC, Feinbaum RL, Ambros V (1993) The C. elegans heterochronic gene lin-4 encodes small RNAs with antisense complementarity to lin-14. Cell 75(5):843–854
Lewis BP, Burge CB, Bartel DP (2005) Conserved seed pairing, often flanked by adenosines, indicates that thousands of human genes are microRNA targets. Cell 120(1):15–20
Li MA, He L (2012) microRNAs as novel regulators of stem cell pluripotency and somatic cell reprogramming. Bioessays 34(8):670–680
Li R, Liang J, Ni S, Zhou T, Qing X, Li H, He W, Chen J, Li F, Zhuang Q, Qin B, Xu J, Li W, Yang J, Gan Y, Qin D, Feng S, Song H, Yang D, Zhang B, Zeng L, Lai L, Esteban MA, Pei D (2010) A mesenchymal-to-epithelial transition initiates and is required for the nuclear reprogramming of mouse fibroblasts. Cell Stem Cell 7(1):51–63
Li Z, Yang C-S, Nakashima K, Rana TM (2011) Small RNA-mediated regulation of iPS cell generation. EMBO J 30(5):823–834
Li T-S, Cheng K, Malliaras K, Smith RR, Zhang Y, Sun B, Matsushita N, Blusztajn A, Terrovitis J, Kusuoka H, Marbán L, Marbán E (2012) Direct comparison of different stem cell types and subpopulations reveals superior paracrine potency and myocardial repair efficacy with cardiosphere-derived cells. J Am Coll Cardiol 59(10):942–953
Li P, Zhu N, Yi B, Wang N, Chen M, You X, Zhao X, Solomides CC, Qin Y, Sun J (2013) MicroRNA-663 regulates human vascular smooth muscle cell phenotypic switch and vascular neointimal formation. Circ Res 113(10):1117–1127
Liang Y, Ridzon D, Wong L, Chen C (2007) Characterization of microRNA expression profiles in normal human tissues. BMC Genomics 8:166
Lin S-L, Chang DC, Chang-Lin S, Lin C-H, Wu DTS, Chen DT, Ying S-Y (2008) Mir-302 reprograms human skin cancer cells into a pluripotent ES-cell-like state. RNA 14(10):2115–2124
Lin S-L, Chang DC, Lin C-H, Ying S-Y, Leu D, Wu DTS (2011) Regulation of somatic cell reprogramming through inducible mir-302 expression. Nucleic Acids Res 39(3):1054–1065
Lipchina I, Elkabetz Y, Hafner M, Sheridan R, Mihailovic A, Tuschl T, Sander C, Studer L, Betel D (2011) Genome-wide identification of microRNA targets in human ES cells reveals a role for miR-302 in modulating BMP response. Genes Dev 25(20):2173–2186
Lipchina I, Studer L, Betel D (2012) The expanding role of miR-302-367 in pluripotency and reprogramming. Cell Cycle 11(8):1517–1523
Liu N, Olson EN (2010) MicroRNA regulatory networks in cardiovascular development. Dev Cell 18(4):510–525
Liu N, Williams AH, Kim Y, McAnally J, Bezprozvannaya S, Sutherland LB, Richardson JA, Bassel-Duby R, Olson EN (2007) An intragenic MEF2-dependent enhancer directs muscle-specific expression of microRNAs 1 and 133. Proc Natl Acad Sci 104(52):20844–20849
Liu N, Bezprozvannaya S, Williams AH, Qi X, Richardson JA, Bassel-Duby R, Olson EN (2008) microRNA-133a regulates cardiomyocyte proliferation and suppresses smooth muscle gene expression in the heart. Genes Dev 22(23):3242–3254
Liu X, Cheng Y, Zhang S, Lin Y, Yang J, Zhang C (2009) A necessary role of miR-221 and miR-222 in vascular smooth muscle cell proliferation and neointimal hyperplasia. Circ Res 104(4):476–487
Lu T-Y, Lin B, Li Y, Arora A, Han L, Cui C, Coronnello C, Sheng Y, Benos PV, Yang L (2013) Overexpression of microRNA-1 promotes cardiomyocyte commitment from human cardiovascular progenitors via suppressing WNT and FGF signaling pathways. J Mol Cell Cardiol 63:146–154
Lund E, Güttinger S, Calado A, Dahlberg JE, Kutay U (2004) Nuclear export of microRNA precursors. Science 303(5654):95–98
Lüningschrör P, Stöcker B, Kaltschmidt B, Kaltschmidt C (2012) miR-290 cluster modulates pluripotency by repressing canonical NF-κB signaling. Stem Cells 30(4):655–664
Lüningschrör P, Hauser S, Kaltschmidt B, Kaltschmidt C (2013) MicroRNAs in pluripotency, reprogramming and cell fate induction. Biochim Biophys Acta 1833(8):1894–1903

Luo Z, Wen G, Wang G, Pu X, Ye S, Xu Q, Wang W, Xiao Q (2013) MicroRNA-200C and -150 play an important role in endothelial cell differentiation and vasculogenesis by targeting transcription repressor ZEB1. Stem Cells 31(9):1749–1762

Mallon BS, Hamilton RS, Kozhich OA, Johnson KR, Fann YC, Rao MS, Robey PG (2014) Comparison of the molecular profiles of human embryonic and induced pluripotent stem cells of isogenic origin. Stem Cell Res 12(2):376–386

Marión RM, Strati K, Li H, Murga M, Blanco R, Ortega S, Fernandez-Capetillo O, Serrano M, Blasco MA (2009) A p53-mediated DNA damage response limits reprogramming to ensure iPS cell genomic integrity. Nature 460(7259):1149–1153

Marks H, Kalkan T, Menafra R, Denissov S, Jones K, Hofemeister H, Nichols J, Kranz A, Stewart AF, Smith A, Stunnenberg HG (2012) The transcriptional and epigenomic foundations of ground state pluripotency. Cell 149(3):590–604

Marson A, Levine SS, Cole MF, Frampton GM, Brambrink T, Johnstone S, Guenther MG, Johnston WK, Wernig M, Newman J, Calabrese JM, Dennis LM, Volkert TL, Gupta S, Love J, Hannett N, Sharp PA, Bartel DP, Jaenisch R, Young RA (2008) Connecting microRNA genes to the core transcriptional regulatory circuitry of embryonic stem cells. Cell 134(3):521–533

Mathonnet G, Fabian MR, Svitkin YV, Parsyan A, Huck L, Murata T, Biffo S, Merrick WC, Darzynkiewicz E, Pillai RS, Filipowicz W, Duchaine TF, Sonenberg N (2007) MicroRNA inhibition of translation initiation in vitro by targeting the cap-binding complex eIF4F. Science 317(5845):1764–1767

Melton C, Judson RL, Blelloch R (2010) Opposing microRNA families regulate self-renewal in mouse embryonic stem cells. Nature 463(7281):621–626

Mitalipov S, Wolf D (2009) Totipotency, pluripotency and nuclear reprogramming. Adv Biochem Eng Biotechnol 114:185–199

Miyoshi N, Ishii H, Nagano H, Haraguchi N, Dewi DL, Kano Y, Nishikawa S, Tanemura M, Mimori K, Tanaka F, Saito T, Nishimura J, Takemasa I, Mizushima T, Ikeda M, Yamamoto H, Sekimoto M, Doki Y, Mori M (2011) Reprogramming of mouse and human cells to pluripotency using mature microRNAs. Cell Stem Cell 8(6):633–638

Muraoka N, Yamakawa H, Miyamoto K, Sadahiro T, Umei T, Isomi M, Nakashima H, Akiyama M, Wada R, Inagawa K, Nishiyama T, Kaneda R, Fukuda T, Takeda S, Tohyama S, Hashimoto H, Kawamura Y, Goshima N, Aeba R, Yamagishi H, Fukuda K, Ieda M (2014) MiR-133 promotes cardiac reprogramming by directly repressing Snai1 and silencing fibroblast signatures. EMBO J 33(14):1565–1581

Murchison EP, Partridge JF, Tam OH, Cheloufi S, Hannon GJ (2005) Characterization of Dicer-deficient murine embryonic stem cells. Proc Natl Acad Sci U S A 102(34):12135–12140

Ng SY, Wong CK, Tsang SY (2010) Differential gene expressions in atrial and ventricular myocytes: insights into the road of applying embryonic stem cell-derived cardiomyocytes for future therapies. Am J Physiol Cell Physiol 299(6):C1234–C1249

Nichols J, Smith A (2009) Naive and primed pluripotent states. Cell Stem Cell 4(6):487–492

Nicoli S, Knyphausen C-P, Zhu LJ, Lakshmanan A, Lawson ND (2012) miR-221 is required for endothelial tip cell behaviors during vascular development. Dev Cell 22(2):418–429

Nishihara T, Zekri L, Braun JE, Izaurralde E (2013) miRISC recruits decapping factors to miRNA targets to enhance their degradation. Nucleic Acids Res 41(18):8692–8705

Nussbaum J, Minami E, Laflamme MA, Virag JAI, Ware CB, Masino A, Muskheli V, Pabon L, Reinecke H, Murry CE (2007) Transplantation of undifferentiated murine embryonic stem cells in the heart: teratoma formation and immune response. FASEB J 21(7):1345–1357

O'Donnell KA, Wentzel EA, Zeller KI, Dang CV, Mendell JT (2005) c-Myc-regulated microRNAs modulate E2F1 expression. Nature 435(7043):839–843

Okano H, Nakamura M, Yoshida K, Okada Y, Tsuji O, Nori S, Ikeda E, Yamanaka S, Miura K (2013) Steps toward safe cell therapy using induced pluripotent stem cells. Circ Res 112(3):523–533

Osafune K, Caron L, Borowiak M, Martinez RJ, Fitz-Gerald CS, Sato Y, Cowan CA, Chien KR, Melton DA (2008) Marked differences in differentiation propensity among human embryonic stem cell lines. Nat Biotechnol 26(3):313–315

Ozsolak F, Poling LL, Wang Z, Liu H, Liu XS, Roeder RG, Zhang X, Song JS, Fisher DE (2008) Chromatin structure analyses identify miRNA promoters. Genes Dev 22(22):3172–3183

Pfaff N, Fiedler J, Holzmann A, Schambach A, Moritz T, Cantz T, Thum T (2011) miRNA screening reveals a new miRNA family stimulating iPS cell generation via regulation of Meox2. EMBO Rep 12(11):1153–1159

Pfaff J, Hennig J, Herzog F, Aebersold R, Sattler M, Niessing D, Meister G (2013) Structural features of Argonaute-GW182 protein interactions. Proc Natl Acad Sci U S A 110(40):E3770–E3779

Porrello ER (2013) microRNAs in cardiac development and regeneration. Clin Sci (Lond) 125(4):151–166

Protze S, Khattak S, Poulet C, Lindemann D, Tanaka EM, Ravens U (2012) A new approach to transcription factor screening for reprogramming of fibroblasts to cardiomyocyte-like cells. J Mol Cell Cardiol 53(3):323–332

Razak SRA, Ueno K, Takayama N, Nariai N, Nagasaki M, Saito R, Koso H, Lai C-Y, Murakami M, Tsuji K, Michiue T, Nakauchi H, Otsu M, Watanabe S (2013) Profiling of microRNA in human and mouse ES and iPS cells reveals overlapping but distinct microRNA expression patterns. PLoS One 8(9), e73532

Reinhart BJ, Slack FJ, Basson M, Pasquinelli AE, Bettinger JC, Rougvie AE, Horvitz HR, Ruvkun G (2000) The 21-nucleotide let-7 RNA regulates developmental timing in Caenorhabditis elegans. Nature 403(6772):901–906

Saini HK, Griffiths-Jones S, Enright AJ (2007) Genomic analysis of human microRNA transcripts. Proc Natl Acad Sci U S A 104(45):17719–17724

Schindler YL, Garske KM, Wang J, Firulli BA, Firulli AB, Poss KD, Yelon D (2014) Hand2 elevates cardiomyocyte production during zebrafish heart development and regeneration. Development 141(16):3112–3122

Schultz J, Lorenz P, Gross G, Ibrahim S, Kunz M (2008) MicroRNA let-7b targets important cell cycle molecules in malignant melanoma cells and interferes with anchorage-independent growth. Cell Res 18(5):549–557

Sinkkonen L, Hugenschmidt T, Berninger P, Gaidatzis D, Mohn F, Artus-Revel CG, Zavolan M, Svoboda P, Filipowicz W (2008) MicroRNAs control de novo DNA methylation through regulation of transcriptional repressors in mouse embryonic stem cells. Nat Struct Mol Biol 15(3):259–267

Sluijter JPG, van Mil A, van Vliet P, Metz CHG, Liu J, Doevendans PA, Goumans M-J (2010) MicroRNA-1 and -499 regulate differentiation and proliferation in human-derived cardiomyocyte progenitor cells. Arterioscler Thromb Vasc Biol 30(4):859–868

Small EM, Sutherland LB, Rajagopalan KN, Wang S, Olson EN (2010) MicroRNA-218 regulates vascular patterning by modulation of Slit-Robo signaling. Circ Res 107(11):1336–1344

Song K, Nam Y-J, Luo X, Qi X, Tan W, Huang GN, Acharya A, Smith CL, Tallquist MD, Neilson EG, Hill JA, Bassel-Duby R, Olson EN (2012) Heart repair by reprogramming non-myocytes with cardiac transcription factors. Nature 485(7400):599–604

Spruce T, Pernaute B, Di-Gregorio A, Cobb BS, Merkenschlager M, Manzanares M, Rodriguez TA (2010) An early developmental role for miRNAs in the maintenance of extraembryonic stem cells in the mouse embryo. Dev Cell 19(2):207–219

Stadtfeld M, Hochedlinger K (2010) Induced pluripotency: history, mechanisms, and applications. Genes Dev 24(20):2239–2263

Suárez Y, Fernández-Hernando C, Yu J, Gerber SA, Harrison KD, Pober JS, Iruela-Arispe ML, Merkenschlager M, Sessa WC (2008) Dicer-dependent endothelial microRNAs are necessary for postnatal angiogenesis. Proc Natl Acad Sci U S A 105(37):14082–14087

Subramanyam D, Lamouille S, Judson RL, Liu JY, Bucay N, Derynck R, Blelloch R (2011) Multiple targets of miR-302 and miR-372 promote reprogramming of human fibroblasts to induced pluripotent stem cells. Nat Biotechnol 29(5):443–448

Suh M-R, Lee Y, Kim JY, Kim S-K, Moon S-H, Lee JY, Cha K-Y, Chung HM, Yoon HS, Moon SY, Kim VN, Kim K-S (2004) Human embryonic stem cells express a unique set of microRNAs. Dev Biol 270(2):488–498

Sun N, Longaker MT, Wu JC (2010) Human iPS cell-based therapy: considerations before clinical applications. Cell Cycle 9(5):880–885
Szczyrba J, Nolte E, Wach S, Kremmer E, Stöhr R, Hartmann A, Wieland W, Wullich B, Grässer FA (2011) Downregulation of Sec23A protein by miRNA-375 in prostate carcinoma. Mol Cancer Res 9(6):791–800
Takahashi K, Yamanaka S (2006) Induction of pluripotent stem cells from mouse embryonic and adult fibroblast cultures by defined factors. Cell 126(4):663–676
Tesar PJ, Chenoweth JG, Brook FA, Davies TJ, Evans EP, Mack DL, Gardner RL, McKay RDG (2007) New cell lines from mouse epiblast share defining features with human embryonic stem cells. Nature 448(7150):196–199
Thiery JP, Acloque H, Huang RYJ, Nieto MA (2009) Epithelial-mesenchymal transitions in development and disease. Cell 139(5):871–890
Thomas M, Lieberman J, Lal A (2010) Desperately seeking microRNA targets. Nat Struct Mol Biol 17(10):1169–1174
Tüfekci KU, Meuwissen RLJ, Genç S (2014) The role of microRNAs in biological processes. Methods Mol Biol 1107:15–31
Urbich C, Kaluza D, Frömel T, Knau A, Bennewitz K, Boon RA, Bonauer A, Doebele C, Boeckel J-N, Hergenreider E, Zeiher AM, Kroll J, Fleming I, Dimmeler S (2012) MicroRNA-27a/b controls endothelial cell repulsion and angiogenesis by targeting semaphorin 6A. Blood 119(6):1607–1616
van Rooij E, Quiat D, Johnson BA, Sutherland LB, Qi X, Richardson JA, Kelm RJ, Olson EN (2009) A family of microRNAs encoded by myosin genes governs myosin expression and muscle performance. Dev Cell 17(5):662–673
van Solingen C, de Boer HC, Bijkerk R, Monge M, van Oeveren-Rietdijk AM, Seghers L, de Vries MR, van der Veer EP, Quax PHA, Rabelink TJ, van Zonneveld AJ (2011) MicroRNA-126 modulates endothelial SDF-1 expression and mobilization of Sca-1(+)/Lin(−) progenitor cells in ischaemia. Cardiovasc Res 92(3):449–455
Wang Y, Medvid R, Melton C, Jaenisch R, Blelloch R (2007) DGCR8 is essential for microRNA biogenesis and silencing of embryonic stem cell self-renewal. Nat Genet 39(3):380–385
Wang Y, Baskerville S, Shenoy A, Babiarz JE, Baehner L, Blelloch R (2008a) Embryonic stem cell-specific microRNAs regulate the G1-S transition and promote rapid proliferation. Nat Genet 40(12):1478–1483
Wang S, Aurora AB, Johnson BA, Qi X, McAnally J, Hill JA, Richardson JA, Bassel-Duby R, Olson EN (2008b) The endothelial-specific microRNA miR-126 governs vascular integrity and angiogenesis. Dev Cell 15(2):261–271
Wang J, He Q, Han C, Gu H, Jin L, Li Q, Mei Y, Wu M (2012) p53-facilitated miR-199a-3p regulates somatic cell reprogramming. Stem Cells 30(7):1405–1413
Wang T, Shi S, Sha H (2013) MicroRNAs in regulation of pluripotency and somatic cell reprogramming: small molecule with big impact. RNA Biol 10(8):1255–1261
Weigert R, Yeung AC, Li J, Donaldson JG (2004) Rab22a regulates the recycling of membrane proteins internalized independently of clathrin. Mol Biol Cell 15(8):3758–3770
Westholm JO, Lai EC (2011) Mirtrons: microRNA biogenesis via splicing. Biochimie 93(11):1897–1904
Wilson KD, Venkatasubrahmanyam S, Jia F, Sun N, Butte AJ, Wu JC (2009) MicroRNA profiling of human-induced pluripotent stem cells. Stem Cells Dev 18(5):749–758
Wilson KD, Hu S, Venkatasubrahmanyam S, Fu J-D, Sun N, Abilez OJ, Baugh JJA, Jia F, Ghosh Z, Li RA, Butte AJ, Wu JC (2010) Dynamic microRNA expression programs during cardiac differentiation of human embryonic stem cells: role for miR-499. Circ Cardiovasc Genet 3(5):426–435
Wilson RC, Tambe A, Kidwell MA, Noland CL, Schneider CP, Doudna JA (2014) Dicer-TRBP complex formation ensures accurate mammalian microRNA biogenesis. Mol Cell 57(3):397–407
Wu F, Yang Z, Li G (2009) Role of specific microRNAs for endothelial function and angiogenesis. Biochem Biophys Res Commun 386(4):549–553

Xie C, Huang H, Sun X, Guo Y, Hamblin M, Ritchie RP, Garcia-Barrio MT, Zhang J, Chen YE (2011) MicroRNA-1 regulates smooth muscle cell differentiation by repressing Kruppel-like factor 4. Stem Cells Dev 20(2):205–210

Xin M, Small EM, Sutherland LB, Qi X, McAnally J, Plato CF, Richardson JA, Bassel-Duby R, Olson EN (2009) MicroRNAs miR-143 and miR-145 modulate cytoskeletal dynamics and responsiveness of smooth muscle cells to injury. Genes Dev 23(18):2166–2178

Xu W, San Lucas A, Wang Z, Liu Y (2014) Identifying microRNA targets in different gene regions. BMC Bioinformatics 15(7):S4

Yang WJ, Yang DD, Na S, Sandusky GE, Zhang Q, Zhao G (2005) Dicer is required for embryonic angiogenesis during mouse development. J Biol Chem 280(10):9330–9335

Yang C-S, Li Z, Rana TM (2011) microRNAs modulate iPS cell generation. RNA 17(8): 1451–1460

Ye D, Wang G, Liu Y, Huang W, Wu M, Zhu S, Jia W, Deng A-M, Liu H, Kang J (2012) MiR-138 promotes induced pluripotent stem cell generation through the regulation of the p53 signaling. Stem Cells 30(8):1645–1654

Yin VP, Lepilina A, Smith A, Poss KD (2012) Regulation of zebrafish heart regeneration by miR-133. Dev Biol 365(2):319–327

Yu F, Li J, Chen H, Fu J, Ray S, Huang S, Zheng H, Ai W (2011) Kruppel-like factor 4 (KLF4) is required for maintenance of breast cancer stem cells and for cell migration and invasion. Oncogene 30(18):2161–2172

Zekri L, Kuzuoğlu-Öztürk D, Izaurralde E (2013) GW182 proteins cause PABP dissociation from silenced miRNA targets in the absence of deadenylation. EMBO J 32(7):1052–1065

Zhang Y, Wang Z, Gemeinhart RA (2013) Progress in microRNA delivery. J Control Release 172(3):962–974

Zhang P, Zheng C, Ye H, Teng Y, Zheng B, Yang X, Zhang J (2014) MicroRNA-365 inhibits vascular smooth muscle cell proliferation through targeting cyclin D1. Int J Med Sci 11(8):765–770

Zhao Y, Samal E, Srivastava D (2005) Serum response factor regulates a muscle-specific microRNA that targets Hand2 during cardiogenesis. Nature 436(7048):214–220

Zhao Y, Ransom JF, Li A, Vedantham V, von Drehle M, Muth AN, Tsuchihashi T, McManus MT, Schwartz RJ, Srivastava D (2007) Dysregulation of cardiogenesis, cardiac conduction, and cell cycle in mice lacking miRNA-1-2. Cell 129(2):303–317

Zhou Q, Gallagher R, Ufret-Vincenty R, Li X, Olson EN, Wang S (2011) Regulation of angiogenesis and choroidal neovascularization by members of microRNA-23~27~24 clusters. Proc Natl Acad Sci U S A 108(20):8287–8292

Chapter 7
Epigenetic Regulation of Cardiac Regeneration

Silvia Agostini, Marco Matteucci, Valentina Casieri, Gaia Papini, and Vincenzo Lionetti

7.1 Self-Regeneration of the Adult Human Heart

A small percentage of cardiomyocyte's proliferation occurs in adult human heart (Bergmann et al. 2009). This extraordinary phenomenon was demonstrated through the measurement of the amount of 14C incorporated into the DNA of cardiomyocytes of individuals born before and after the nuclear bomb tests and the comparison with the levels of atmospheric 14C. Interestingly, the detectable pool of new cardiomyocytes ranged from 1 % per year in young adults to 0.45 % in the elderly. While these renewal rates cannot support the endogenous myocardial self-regeneration in response to injury, they have nonetheless encouraged the study of any regulatory mechanism able to enhance it. Heart failure is an heterogeneous disease and recent scientific reports have demonstrated that the improvement of intercellular cross-talk is essential to foster a more effective adaptive response to injury (Tirziu et al. 2010). Less than a third of the total cell number of the adult myocardium consists of cardiomyocytes, which communicate with a broad pool of additional cell types, such as smooth muscle cells, endothelial cells, fibroblasts, mast cells, immune system-related cells and progenitor cells (Bu et al. 2009). The myocardial homeostasis implies different cell-to-cell and cell-to-matrix connections, which form a mature and self-regulated functional unit (Ausoni and Sartore 2009). These distinct cell pools also interact via a variety of soluble paracrine, autocrine and endocrine factors, which require a stable apparatus of gene expression (Lionetti et al. 2010a, b). Recent studies of developmental biology and integrative physiology have highlighted the role played by epigenetic modifications as common regulatory pathway of cell function and threshold of cell turnover in adult's heart (Romano and Lionetti 2013). The epigenetic signals in fact activate, maintain, and change the cardiac transcriptional status

S. Agostini • M. Matteucci • V. Casieri • G. Papini • V. Lionetti, M.D., Ph.D. (✉)
Laboratory of Medical Science, Institute of Life Sciences, Scuola Superiore Sant'Anna, Via G. Moruzzi, 1, Pisa, Italy
e-mail: v.lionetti@sssup.it

R. Madonna (ed.), *Stem Cells and Cardiac Regeneration*, Stem Cell Biology and Regenerative Medicine, DOI 10.1007/978-3-319-25427-2_7

underlying the tissue ability to tolerate the microenvironmental stress without losing functional coherence. All epigenetic pathways remodel chromatin and ensure the maintenance of cardiac homeostasis during several cell divisions and across generations without altering the DNA sequence. Lack of gene function and the cell death follow the failure of balancing epigenetic signals. Specific signaling pathways may drive the phenotype of cardiac resident cells at transcriptional and post-transcriptional levels in healthy and damaged hearts (Lionetti and Ventura 2013a, b). In particular, the following pathways play a key role in the self-regulation of the epigenetic state of adult cardiac cells: DNA methylation, histone modifications, and RNA-based modulation (Issa and Baylin 1996).

7.2 The Epigenetic Memory of Cardiac Cells: A Prelude to Myocardial Regeneration

The epigenetic memory relies on the ability of cardiac cells to translate environmental cues in adaptive response and it remains unchanged after several cell divisions. High levels of DNA methylation lead to maintenance of cell epigenetic memory (Sanchez-Freire et al. 2014). In addition, histone modifications support epigenetic inheritance mechanisms of adult cardiac cells during transient or persistent exposure to different microenvironments (Schlingman et al. 2014). Histone modifications, which determine the structure and function of the chromatin, transmit stable epigenetic information between proliferating cells. Finally, pool of regulatory small noncoding RNAs (ncRNAs) regulate the expression of factors leading to histone modifications, such as the Polycomb-group proteins (Mathiyalagan et al. 2014).

7.2.1 DNA Methylation

In cardiac cells the magnitude of DNA methylation at the C5 position of CpG islands is a balance of two biological processes: (1) de novo methylation or (2) maintenance of methylation during cellular DNA replication. Based on recent findings, the maintenance of methylation assumes a relevant biological role as DNA replication occurs in cardiomyocytes during adulthood (Bergmann et al. 2009). Cardiac DNA demethylation is specifically regulated by Ten-eleven translocation methylcytosine dioxygenase enzymes (TET) that oxidizes the methyl group of specific methylcytosines (Kinney et al. 2011).

The balance between DNA methylation or demethylation affects the expression and function of genes which regulate cardiac function. For example, low levels of circulating methylated Long Interspersed Nucleotide Elements-1 (LINE-1) gene, a biomarker of systemic DNA methylation (Yang et al. 2004), predicts intolerance to myocardial ischemia in elderly individuals (Baccarelli et al. 2010). Moreover, DNA hypermethylation even limits contractile function of rodent cardiomyocytes chronically exposed to norepinephrine-induced hypertrophy (Xiao et al. 2014). Different levels of DNA methylation in normal and failing hearts suggest that changes of epigenomic patterns affect the ability to predict the development of heart failure in humans (Movassagh et al. 2011). It is conceivable that high levels of DNA methylation represent an adaptive mechanism underlying the response to injury by adult cardiomyocytes.

Despite this fascinating hypothesis, DNA methylation does not affect tyrosine kinase-type cell surface receptor HER3 and Homeobox B13 genes, which play a key role in the expression of adaptive cardiac features in response to sustained contractile dysfunction (Haas et al. 2013).

7.2.2 Histone Modifications

Histones are basic proteins of the chromatin (H1, H2A and B, H3 and H4) that form the nucleosome after binding DNA. Histone modifications include the following enzymatic processes: acetylation (Taylor and Liew 1976), phosphorylation (Akhtar and Itzhaki 1977; Liew and Sole 1978), methylation (Kaneda et al. 2009), ADP-ribosylation (Martinez-Zamudio and Ha 2012), biotinylation (Kuroishi et al. 2011), ubiquitination (Zhang 2003) and sumoylation (Wang and Dasso 2009).

The balance between histone acetyltransferases (HATs) or deacetylases (HDACs) modulates DNA transcription in adult cardiac cells of animal models and humans (Miyamoto et al. 2006; Lee et al. 2007; Hariharan et al. 2010). In particular, nuclear HATs play a key role in cardiac hypertrophy (Qiao et al. 2014) and dilation (Miyamoto et al. 2006). However, loss of HATs activity causes the death of embryos due to serious cardiac congenital defects in p300 knock-in mice, where embryonal stem cells do not respond to cardiomyogenic factors (i.e.: BMP-2) (Shikama et al. 2003).

In 2010, we were one of the first to show that the early pharmacological inhibition of cardiac class I HDACs by intramyocardial injection of low dose of hyaluronan mixed estersof butyric and retinoic acid (HBR) induces the expression of paracrine factors that enhance angiogenesis (i.e.: vascular endothelial growth factor, VEGF) and survival/proliferation (i.e.: hepatocyte growth factor, HGF) of adult rat cardiomyocytes bordering the infarct zone. All these changes have preserved the cardiac function in a rodent model of myocardial infarction (Lionetti et al. 2010a, b). Other investigators have confirmed our data in vivo (Zhang et al. 2012). Preliminary evidences provided by us have also shown that HBR hampers cell proliferation and migration of cardiac fibroblasts, attenuates the differentiation to myofibroblasts and inhibits collagen expression (Cavallini et al. 2011). These evidences support the hypothesis that the inhibition of class I HDAC) differently affects the myocardial epigenetic memory depending on the cell type.

Histone H3 phosphorylation at serine-10, a different histone modification, is detectable in proliferating cardiomyocytes of heart exposed to unloading conditions (Canseco et al. 2015) or stimulated with growth factors (Illi et al. 2005). In addition,

Histone H3 Ser-10 phosphorylation leads to transcription of Mef2, a transcription factor that induces cell growth (Awad et al. 2013).

Mono-, di- or tri-methylation of arginine or lysine residues of histone H3 and H4 leads to active or repressed states of chromatin in murine adult cardiac cells (Chaturvedi et al. 2014). Loss of methylation at lysine K4 of histone H3 (H3K4) increases intracellular calcium levels resulting in improved contractile function (Stein et al. 2011). The role of histone biotinylation, ADP-ribosylation and sumoylation in myocardial adaptive response to stressors is still unknown.

7.2.3 RNA-Based Epigenetic Modulators

MicroRNAs (miR) are small non-coding RNAs, long about 22 nucleotides, that degrade the complementary mRNA after binding the RNA-induced silencing complex (RISC) (Brennecke et al. 2005). To date, miRNAs play a role as pathophysiological hallmark of cardiac disease/repair at post-trascriptional level. For example, high circulating levels of miR-1, 21, 133a, 208 and 499 are detectable after myocardial injury (Zile et al. 2011).

In addition, plasma levels of cardiac muscle-enriched miR (i.e.: miR-133a and 208a) increase in patients with coronary artery disease (Fichtlscherer et al. 2010). The profile of circulating miRNAs may be helpful to identify cardiovascular patients even at earlier and later stages of disease. They are also helpful to track the short- and long-term effects of different regenerative approach. In fact, high levels of miR-323-3p and -652 are detected in untreated patients affected by acute coronary syndrome compared to healthy controls up to 4 months post-hospitalization (Pilbrow et al. 2014) and significantly improve the risk stratification in combination with established biomarkers of cardiac dysfunction, such as high levels of N-terminal of the prohormone brain natriuretic peptide (NT-proBNP) and low left ventricular ejection fraction (Pilbrow et al. 2014). Recent studies even highlighted the role of miRNAs as epigenetic paracrine mediators of cell-to-cell/cell-to-matrix interactions in adult heart. Cardioprotective and proangiogenic miRNAs are released into exosomes, nanosized extracellular vesicles, which are detectable in myocardial interstitium and have a strong therapeutic potential (Cervio et al. 2015).

Long non-coding RNAs (LncRNAs), long about 200 nucleotides, play a role in regulating cardiac physiological traits at both transcriptional and post-transcriptional levels, even if they lack significant protein-coding potential (Guttman et al. 2009). Despite LncRNAs expression being affected by cardiovascular disorders (Ounzain et al. 2015), their epigenetic role in mediating myocardial regeneration is still not well defined. Emerging evidences suggest that LncRNAs modulate the expression of miRNAs during hypertrophic response of adult cardiomyocytes to stressors, as angiotensin II (Wang et al. 2014).

Deregulated epigenetic patterns through LncRNAs expression underlies the lack of angiogenic ability of coronary endothelial cells in response to hypoxia. For example, lower levels of metastasis-associated lung adenocarcinoma transcript 1

(MALAT1), a lncRNA highly expressed in hypoxic adult endothelial cells, inhibit endothelial cell proliferation and VEGF-dependent vessel growth (Michalik et al. 2014). A recent study demonstrated that lncRNAs have a differential abundance in exosomes, indicating a selective loading by producing cells (Gezer et al. 2014).

7.3 Epigenetic Modifications in Regenerating Myocardium

Epigenetic modifications are emerging as endogenous mechanisms sustaining cardiac self-renewing properties and are therapeutic candidates for regeneration of heart failure.

7.3.1 DNA Methylation

Leferovich et al. (2001) have observed a high cardiac regenerative potential in MRL/MpJ mice, which show cells arrested in the G2/M phase and high activity of matrix metalloproteinases. In particular, the heart of adult MRL/MpJ mice shows high levels of DNA methylation and retains embryonic features (Górnikiewicz et al. 2013). However, the heart of these mice does not heal after myocardial infarction (Cimini et al. 2008) and I/R injury (Abdullah et al. 2005). This observation suggests that the microenvironment promotes the switch on DNA methylation peak from embryonic to adult profile therefore limiting the regenerative potential of the myocardium. This model may be helpful to detect the genes responsible of the low regenerative capacity of adult heart. Recent studies have demonstrated that Notch-responsive promoters, which support cardiomyocytes proliferation in zebrafish heart (Zhao et al. 2014), show higher levels of permanent CpG DNA methylation in adult murine cardiomyocytes of infarcted hearts compared to healthy ones (Felician et al. 2014). Notch-responsive promoters are permanently silenced by DNA methylation.

Conversely, another study has shown that the pharmacological inhibition of DNA methylation in rodent stem cells induces protein expression of homeobox protein Nkx2.5, transcription factor GATA binding protein 4 (GATA4) and cardiac troponin T which trigger the differentiation to cardiac lineage (Naeem et al. 2013). Therefore, the modulators of levels of DNA methylation are therapeutic candidate to enhance myocardial regeneration.

7.3.2 Histone Modifications

HATs activation increases acetylation of lysine K9 and K14 of histone H3 at physiological level, reduces levels of DNA CpG methylation, and recovers the ability of mesenchymal stem cells to proliferate and differentiate to cardiac lineage in the presence of oxidative stress (Vecellio et al. 2014).

Conversely, the inhibition of HDACs induces the proliferation of adult rodent cardiomyocytes both in vitro and in vivo (Majumdar et al. 2012). In small and large animal models of heart failure, myocardial histone acetylation increase the expression and release of paracrine factors, such as the hepatocyte growth factor (HGF) (Iwasaki et al. 2005) or the vascular endothelial growth factor (VEGF) (Tao et al. 2011).

Some growth factors may in turn activate HDACs. For example, VEGF play an epigenetic role by inducing the degradation of HDAC7 via phospholipase C gamma-inositol-1,4,5-trisphosphate kinase signal pathway, which prevents the HDAC7-mediated inhibition of cell proliferation, as observed in mature endothelial cells (Margariti et al. 2010).

Recent studies have highlighted the epigenetic role of other important paracrine factors involved in myocardial angiogenesis. Li et al. (2014) have demonstrated that carbon monoxide enhances the levels of histone acetylation, which in turn improves endothelial cell migration, and increases the angiogenic ability of human endothelial cells (Lionetti et al. 2010a, b; Agostini et al. 2015) following the treatment with inhibitors of class I HDACs. These data refute previous in vitro studies demonstrating that the maintenance of HDAC activity induces angiogenesis (Kim et al. 2001; Mottet et al. 2007).

Furthermore, Mezentseva et al. (2013) have demonstrated in vitro that low levels of methylation at lysine9 of histone H3 leads to reprogramming of bone marrow stem cells towards a cardiac lineage.

These studies encourage the development of novel and safer histone modulators also to optimize the cardiac differentiation of circulating bone marrow-derived stem cells engrafted in the failing heart.

7.3.3 RNA-Based Epigenetic Modulators

The cardiac delivery of miR-590 and -199 promoted cell cycle re-entry of adult cardiomyocytes and enhanced cardiomyocyte proliferation in infarcted murine heart (Eulalio et al. 2012). These data are in accord with previous studies suggesting that endogenous miRNAs act as endogenous regulators of cell reprogramming and as therapeutic targets in the setting of novel avenue in cell-free cardiac regeneration. However, miRNAs even may exert negative effect on cell function.

Bonauer et al. (2009) have shown that high levels of miR-92a inhibits the formation of new blood vessels, which play a relevant role in the functional recovery of murine infarcted heart. In light of this evidence, the synthesis of miRNA antagonists may be a helpful tool to promote myocardial regeneration. In fact, treatment with selective LNA-modified anti-miR-15 prevented loss of hypoxic cardiomyocytes, reduced infarct scar size and preserved cardiac function in murine infarcted heart (Hullinger et al. 2012). Interestingly, stable myocardial down-regulation of miR-24 enhances angiogenesis and blood perfusion in the myocardium surrounding the infarct area and improves cardiac function despite promoting apoptosis of fibroblasts and cardiomyocytes (Meloni et al. 2013).

To date, it remains unclear whether the modulation of single miR-dependent pathway is sufficient to trigger cardiomyocyte's proliferation in regenerating the adult infarcted heart. The selective inhibition of miRNA-15 even increases the rate of proliferation of adult cardiomyocytes and significantly improves the cardiac function of infarcted murine heart (Porrello et al. 2013). Conversely, other investigators found that an effective proliferation of adult cardiomyocytes require the combined action of several miRNAs, such as miR17-92 cluster (Chen et al. 2013). Jayawardena et al. (Jayawardena et al. 2012) have demonstrated in mice that a cocktail of miR-1, -133, -208 and -499 restores the post-ischemic loss of cardiomyocyte's pool with a population of fibroblast-derived cardiomyocytes due to gene reprogramming. All these findings suggest that mir-crine mechanisms hold great promise as therapeutic candidates for the development of personalized myocardial regeneration.

Regulatory miRNAs are naturally released by cardiomyocytes, endothelial cells, fibroblasts or cardiac/endothelial progenitor cells, as demonstrated in adult mice (Brás-Rosário et al. 2013). Exosomes containing miRNAs are released into the extracellular microenvironment by adult cardiomyocytes (Wang et al. 2014), cardiac fibroblasts (Bang et al. 2014), endothelial cells (Ong et al. 2014) and cardiac progenitor cells (Vrijsen et al. 2010).

Exosomes secreted by human cardiac progenitor cells (CPCs) contain higher amounts of miRNA210, 132 and 146a-3p compared with human cardiac fibroblasts (Barile et al. 2014). Each miRNA acts on specific pathways; miRNA210 inhibits cardiomyocyte's apoptosis through the down regulation of ephrin A3 and protein-tyrosine phosphatase 1B (PTP1B) expression, while miRNA132 increases the angiogenic ability of mature endothelial cells through the down regulation of RasGAP-p120 expression. The single intramyocardial injection of CPCs-derived exosomes hampers cardiac remodeling and preserves left ventricular ejection fraction of infarcted rat heart in a dose-dependent manner. Conversely, fibroblasts-derived exosomes did not exert any cardioprotective effects. Other investigators have found similar findings in a murine model of heart failure (Ibrahim et al. 2014).

Some investigators are developing new methods to increase the endogenous release of cardioprotective exosomes containing miRNAs. Original study found that high levels of hypoxia-inducible factor-1 (HIF-1), a transcription factor that protects against ischemia and highly expressed in human failing heart (Lionetti et al. 2014), increase the release of exosomes containing miRNA126 and 210 (Ong et al. 2014). Therefore, the exosomes have the potential to circumvent many limitations of stem cells transplantation for therapeutic applications in cardiac regenerative medicine.

Recently, it has been shown that exosomes deliver LncRNAs (Gezer et al. 2014). To date, a few studies have characterized the regenerative potential of LncRNAs in the adult heart. For example, the LncRNA steroid receptor RNA activator 1 (SRA1) regulates the expression of myogenic differentiation 1 (MyoD1) (Caretti et al. 2006) and is essential for myocardial function. Further investigations should be encouraged to better address the therapeutic potential of exosomes containing LncRNAs.

7.4 Conclusions and Perspectives

Studies conducted so far have provided convincing experimental evidences that different epigenetic mechanisms underlie the changes of the cardiac physiological traits during myocardial regeneration. The pleiotropic non-invasive modulation of the epigenetic threshold of resident cardiac cells by drugs or diet is a frontier of investigation that should be encouraged in order to overcome the limitations that hinder an efficient structural and functional recovery of the adult heart in an epigenetic manner. In fact, emerging scientific evidences have shown that it is possible to modulate the cardiac epigenome and to induce cardiac benefits by the regular intake of lower doses of active plant compounds, such as barley beta-glucan (Agostini et al. 2015), or by the administration of selected exosomes (Barile et al. 2014). To best of our knowledge, the cardiac repair following the administration of the potential epigenetic modulators will be more effective at lower doses and mainly focused to histone or RNA-based modifications. Further translational investigations should be conducted in large animal models of heart failure and humans to better address dose, timing and route of administration.

References

Abdullah I, Lepore JJ, Epstein JA, Parmacek MS, Gruber PJ (2005) MRL mice fail to heal the heart in response to ischemia reperfusion injury. Wound Repair Regen 13:205–208

Agostini S, Chiavacci E, Matteucci M, Torelli M, Pitto L, Lionetti V (2015) Barley beta-glucan promotes MnSOD expression and enhances angiogenesis under oxidative microenvironment. J Cell Mol Med 19:227–238

Akhtar RA, Itzhaki S (1977) Studies in vitro of the effects of adenosine 3′:5′-cyclic monophosphate on the phosphorylation of nuclear proteins in isolated rat heart nuclei. Biochem J 161:487–494

Ausoni S, Sartore S (2009) The cardiovascular unit as a dynamic player in disease and regeneration. Trends Mol Med 15:543–552

Awad S, Kunhi M, Little GH, Bai Y, An W, Bers D, Kedes L, Poizat C (2013) Nuclear CaMKII enhances histone H3 phosphorylation and remodels chromatin during cardiac hypertrophy. Nucleic Acids Res 41:7656–7672

Baccarelli A, Wright R, Bollati V, Litonjua A, Zanobetti A, Tarantini L, Sparrow D, Vokonas P, Schwartz J (2010) Ischemic heart disease and stroke in relation to blood DNA methylation. Epidemiology 21:819–828

Bang C, Batkai S, Dangwal S, Gupta SK, Foinquinos A, Holzmann A, Just A, Remke J, Zimmer K, Zeug A, Ponimaskin E, Schmiedl A, Yin X, Mayr M, Halder R, Fischer A, Engelhardt S, Wei Y, Schober A, Fiedler J, Thum T (2014) Cardiac fibroblast-derived microRNA passenger strand enriched exosomes mediate cardiomyocyte hypertrophy. J Clin Invest 124:2136–2146

Barile L, Lionetti V, Cervio E, Matteucci M, Gherghiceanu M, Popescu LM, Torre T, Siclari F, Moccetti T, Vassalli G (2014) Extracellular vesicles from human cardiac progenitor cells inhibit cardiomyocyte apoptosis and improve cardiac function after myocardial infarction. Cardiovasc Res 103:530–541

Bergmann O, Bhardwaj RD, Bernard S, Zdunek S, Barnabé-Heider F, Walsh S, Zupicich J, Alkass K, Buchholz BA, Druid H, Jovinge S, Frisén J (2009) Evidence for cardiomyocyte renewal in humans. Science 324:98–102

Bonauer A, Carmona G, Iwasaki M, Mione M, Koyanagi M, Fischer A, Burchfield J, Fox H, Doebele C, Ohtani K, Chavakis E, Potente M, Tjwa M, Urbich C, Zeiher AM, Dimmeler S (2009) MicroRNA-92a controls angiogenesis and functional recovery of ischemic tissues in mice. Science 324:1710–1713

Brás-Rosário L, Matsuda A, Pinheiro AI, Gardner R, Lopes T, Amaral A, Gama-Carvalho M (2013) Expression profile of microRNAs regulating proliferation and differentiation in mouse adult cardiac stem cells. PLoS One 8:e63041

Brennecke J, Stark A, Russell RB, Cohen SM (2005) Principles of microRNA-target recognition. PLoS Biol 3, e85

Bu L, Jiang X, Martin-Puig S, Caron L, Zhu S, Shao Y, Roberts DJ, Huang PL, Domian IJ, Chien KR (2009) Human ISL1 heart progenitors generate diverse multipotent cardiovascular cell lineages. Nature 460:113–117

Canseco DC, Kimura W, Garg S, Mukherjee S, Bhattacharya S, Abdisalaam S, Das S, Asaithamby A, Mammen PP, Sadek HA (2015) Human ventricular unloading induces cardiomyocyte proliferation. J Am Coll Cardiol 65:892–900

Caretti G, Schiltz RL, Dilworth FJ, Di Padova M, Zhao P, Ogryzko V, Fuller-Pace FV, Hoffman EP, Tapscott SJ, Sartorelli V (2006) The RNA helicases p68/p72 and the noncoding RNA SRA are coregulators of MyoD and skeletal muscle differentiation. Dev Cell 11:547–560

Cavallini C, Tassinari R, Bonavita F, Lionetti V, Ventura C (2011) Hyaluronan mixed esters of butyric and retinoic acid act transcriptionally on cardiac fibroblasts decreasing myocardial scarring in infarcted hearts. FASEB J 25: 2:2

Cervio E, Barile L, Moccetti T, Vassalli G (2015) Exosomes for intramyocardial intercellular communication. Stem Cells Int 2015:482171

Chaturvedi P, Kalani A, Givvimani S, Kamat PK, Familtseva A, Tyagi SC (2014) Differential regulation of DNA methylation versus histone acetylation in cardiomyocytes during HHcy in vitro and in vivo: an epigenetic mechanism. Physiol Genomics 46:245–255

Chen J, Huang ZP, Seok HY, Ding J, Kataoka M, Zhang Z, Hu X, Wang G, Lin Z, Wang S, Pu WT, Liao R, Wang DZ (2013) mir-17-92 cluster is required for and sufficient to induce cardiomyocyte proliferation in postnatal and adult hearts. Circ Res 112:1557–1566

Cimini M, Fazel S, Fujii H, Zhou S, Tang G, Weisel RD, Li RK (2008) The MRL mouse heart does not recover cardiac function after a myocardial infarction. Cardiovasc Pathol 17:32–39

Eulalio A, Mano M, Dal Ferro M, Zentilin L, Sinagra G, Zacchigna S, Giacca M (2012) Functional screening identifies miRNAs inducing cardiac regeneration. Nature 492:376–381

Felician G, Collesi C, Lusic M, Martinelli V, Ferro MD, Zentilin L, Zacchigna S, Giacca M (2014) Epigenetic modification at notch responsive promoters blunts efficacy of inducing notch pathway reactivation after myocardial infarction. Circ Res 115:636–649

Fichtlscherer S, De Rosa S, Fox H, Schwietz T, Fischer A, Liebetrau C, Weber M, Hamm CW, Röxe T, Müller-Ardogan M, Bonauer A, Zeiher AM, Dimmeler S (2010) Circulating microRNAs in patients with coronary artery disease. Circ Res 107:677–684

Gezer U, Özgür E, Cetinkaya M, Isin M, Dalay N (2014) Long non-coding RNAs with low expression levels in cells are enriched in secreted exosomes. Cell Biol Int 38:1076–1079

Górnikiewicz B, Ronowicz A, Podolak J, Madanecki P, Stanisławska-Sachadyn A, Sachadyn P (2013) Epigenetic basis of regeneration: analysis of genomic DNA methylation profiles in the MRL/MpJ mouse. DNA Res 20:605–621

Guttman M, Amit I, Garber M, French C, Lin MF, Feldser D, Huarte M, Zuk O, Carey BW, Cassady JP, Cabili MN, Jaenisch R, Mikkelsen TS, Jacks T, Hacohen N, Bernstein BE, Kellis M, Regev A, Rinn JL, Lander ES (2009) Chromatin signature reveals over a thousand highly conserved large non-coding RNAs in mammals. Nature 458:223–227

Haas J, Frese KS, Park YJ, Keller A, Vogel B, Lindroth AM, Weichenhan D, Franke J, Fischer S, Bauer A, Marquart S, Sedaghat-Hamedani F, Kayvanpour E, Köhler D, Wolf NM, Hassel S, Nietsch R, Wieland T, Ehlermann P, Schultz JH, Dösch A, Mereles D, Hardt S, Backs J, Hoheisel JD, Plass C, Katus HA, Meder B (2013) Alterations in cardiac DNA methylation in human dilated cardiomyopathy. EMBO Mol Med 5:413–429

Hariharan N, Maejima Y, Nakae J, Paik J, Depinho RA, Sadoshima J (2010) Deacetylation of FoxO by Sirt1 plays an essential role in mediating starvation-induced autophagy in cardiac myocytes. Circ Res 107:1470–1482

Hullinger TG, Montgomery RL, Seto AG, Dickinson BA, Semus HM, Lynch JM, Dalby CM, Robinson K, Stack C, Latimer PA, Hare JM, Olson EN, van Rooij E (2012) Inhibition of miR-1 5 protects against cardiac ischemic injury. Circ Res 110:71–81

Ibrahim AG, Cheng K, Marbán E (2014) Exosomes as critical agents of cardiac regeneration triggered by cell therapy. Stem Cell Rep 2:606–619

Illi B, Scopece A, Nanni S, Farsetti A, Morgante L, Biglioli P, Capogrossi MC, Gaetano C (2005) Epigenetic histone modification and cardiovascular lineage programming in mouse embryonic stem cells exposed to laminar shear stress. Circ Res 96:501–508

Issa JP, Baylin SB (1996) Epigenetics and human disease. Nat Med 2:281–282

Iwasaki M, Adachi Y, Nishiue T, Minamino K, Suzuki Y, Zhang Y, Nakano K, Koike Y, Wang J, Mukaide H, Taketani S, Yuasa F, Tsubouchi H, Gohda E, Iwasaka T, Ikehara S (2005) Hepatocyte growth factor delivered by ultrasound-mediated destruction of microbubbles induces proliferation of cardiomyocytes and amelioration of left ventricular contractile function in Doxorubicin-induced cardiomyopathy. Stem Cells 23:1589–1597

Jayawardena TM, Egemnazarov B, Finch EA, Zhang L, Payne JA, Pandya K, Zhang Z, Rosenberg P, Mirotsou M, Dzau VJ (2012) MicroRNA-mediated in vitro and in vivo direct reprogramming of cardiac fibroblasts to cardiomyocytes. Circ Res 110:1465–1473

Kaneda R, Takada S, Yamashita Y, Choi YL, Nonaka-Sarukawa M, Soda M, Misawa Y, Isomura T, Shimada K, Mano H (2009) Genome-wide histone methylation profile for heart failure. Genes Cells 14:69–77

Kim MS, Kwon HJ, Lee YM, Baek JH, Jang JE, Lee SW, Moon EJ, Kim HS, Lee SK, Chung HY, Kim CW, Kim KW (2001) Histone deacetylases induce angiogenesis by negative regulation of tumor suppressor genes. Nat Med 7:437–443

Kinney SM, Chin HG, Vaisvila R, Bitinaite J, Zheng Y, Estève PO, Feng S, Stroud H, Jacobsen SE, Pradhan S (2011) Tissue-specific distribution and dynamic changes of 5-hydroxymethylcytosine in mammalian genomes. J Biol Chem 286:24685–24693

Kuroishi T, Rios-Avila L, Pestinger V, Wijeratne SS, Zempleni J (2011) Biotinylation is a natural, albeit rare, modification of human histones. Mol Genet Metab 104:537–545

Lee TM, Lin MS, Chang NC (2007) Inhibition of histone deacetylase on ventricular remodeling in infarcted rats. Am J Physiol Heart Circ Physiol 293:H968–H977

Leferovich JM, Bedelbaeva K, Samulewicz S, Zhang XM, Zwas D, Lankford EB, Heber-Katz E (2001) Heart regeneration in adult MRL mice. Proc Natl Acad Sci U S A 98:9830–9835

Li M, Gallo D, Csizmadia E, Otterbein LE, Wegiel B (2014) Carbon monoxide induces chromatin remodelling to facilitate endothelial cell migration. Thromb Haemost 111:951–959

Liew CC, Sole MJ (1978) Studies of nuclear proteins in the heart of the cardiomyopathic Syrian hamster—phosphorylation of histones. J Mol Cell Cardiol 10:847–855

Lionetti V, Ventura C (2013a) New chemicals drive mesenchymal stem cells toward cardiac cells. Recent Pat Regen Med 3:47–55

Lionetti V, Ventura C (2013b) Regenerative medicine approach to repair the failing heart. Vascul Pharmacol 58:159–163

Lionetti V, Bianchi G, Recchia FA, Ventura C (2010a) Control of autocrine and paracrine myocardial signals: an emerging therapeutic strategy in heart failure. Heart Fail Rev 15:531–542

Lionetti V, Cantoni S, Cavallini C, Bianchi F, Valente S, Frascari I, Olivi E, Aquaro GD, Bonavita F, Scarlata I, Maioli M, Vaccari V, Tassinari R, Bartoli A, Recchia FA, Pasquinelli G, Ventura C (2010b) Hyaluronan mixed esters of butyric and retinoic acid affording myocardial survival and repair without stem cell transplantation. J Biol Chem 285:9949–9961

Lionetti V, Matteucci M, Ribezzo M, Di Silvestre D, Brambilla F, Agostini S, Mauri P, Padeletti L, Pingitore A, Delsedime L, Rinaldi M, Recchia FA, Pucci A (2014) Regional mapping of myocardial hibernation phenotype in idiopathic end-stage dilated cardiomyopathy. J Cell Mol Med 18:396–414

Majumdar G, Adris P, Bhargava N, Chen H, Raghow R (2012) Pan-histone deacetylase inhibitors regulate signaling pathways involved in proliferative and pro-inflammatory mechanisms in H9c2 cells. BMC Genomics 13:709

Margariti A, Zampetaki A, Xiao Q, Zhou B, Karamariti E, Martin D, Yin X, Mayr M, Li H, Zhang Z, De Falco E, Hu Y, Cockerill G, Xu Q, Zeng L (2010) Histone deacetylase 7 controls endothelial cell growth through modulation of beta-catenin. Circ Res 106:1202–1211

Martinez-Zamudio R, Ha HC (2012) Histone ADP-ribosylation facilitates gene transcription by directly remodeling nucleosomes. Mol Cell Biol 32:2490–2502

Mathiyalagan P, Okabe J, Chang L, Su Y, Du XJ, El-Osta A (2014) The primary microRNA-208b interacts with Polycomb-group protein, Ezh2, to regulate gene expression in the heart. Nucleic Acids Res 42:790–803

Meloni M, Marchetti M, Garner K, Littlejohns B, Sala-Newby G, Xenophontos N, Floris I, Suleiman MS, Madeddu P, Caporali A, Emanueli C (2013) Local inhibition of microRNA-24 improves reparative angiogenesis and left ventricle remodeling and function in mice with myocardial infarction. Mol Ther 21:1390–1402

Mezentseva NV, Yang J, Kaur K, Iaffaldano G, Rémond MC, Eisenberg CA, Eisenberg LM (2013) The histone methyltransferase inhibitor BIX01294 enhances the cardiac potential of bone marrow cells. Stem Cells Dev 22:654–667

Michalik KM, You X, Manavski Y, Doddaballapur A, Zörnig M, Braun T, John D, Ponomareva Y, Chen W, Uchida S, Boon RA, Dimmeler S (2014) Long noncoding RNA MALAT1 regulates endothelial cell function and vessel growth. Circ Res 114:1389–1397

Miyamoto S, Kawamura T, Morimoto T, Ono K, Wada H, Kawase Y, Matsumori A, Nishio R, Kita T, Hasegawa K (2006) Histone acetyltransferase activity of p300 is required for the promotion of left ventricular remodeling after myocardial infarction in adult mice in vivo. Circulation 113:679–690

Mottet D, Bellahcène A, Pirotte S, Waltregny D, Deroanne C, Lamour V, Lidereau R, Castronovo V (2007) Histone deacetylase 7 silencing alters endothelial cell migration, a key step in angiogenesis. Circ Res 101:1237–1246

Movassagh M, Choy MK, Knowles DA, Cordeddu L, Haider S, Down T, Siggens L, Vujic A, Simeoni I, Penkett C, Goddard M, Lio P, Bennett MR, Foo RS (2011) Distinct epigenomic features in end-stage failing human hearts. Circulation 124:2411–2422

Naeem N, Haneef K, Kabir N, Iqbal H, Jamall S, Salim A (2013) DNA methylation inhibitors, 5 azacytidine and zebularine potentiate the transdifferentiation of rat bone marrow mesenchymal stem cells into cardiomyocytes. Cardiovasc Ther 31:201–209

Ong SG, Lee WH, Huang M, Dey D, Kodo K, Sanchez-Freire V, Gold JD, Wu JC (2014) Cross talk of combined gene and cell therapy in ischemic heart disease: role of exosomal microRNA transfer. Circulation 130:S60–S69

Ounzain S, Micheletti R, Beckmann T, Schroen B, Alexanian M, Pezzuto I, Crippa S, Nemir M, Sarre A, Johnson R, Dauvillier J, Burdet F, Ibberson M, Guigó R, Xenarios I, Heymans S, Pedrazzini T (2015) Genome-wide profiling of the cardiac transcriptome after myocardial infarction identifies novel heart-specific long non-coding RNAs. Eur Heart J 36:353–368a

Pilbrow AP, Cordeddu L, Cameron VA, Frampton CM, Troughton RW, Doughty RN, Whalley GA, Ellis CJ, Yandle TG, Richards AM, Foo RS (2014) Circulating miR-323-3p and miR-652: candidate markers for the presence and progression of acute coronary syndromes. Int J Cardiol 176:375–385

Porrello ER, Mahmoud AI, Simpson E, Johnson BA, Grinsfelder D, Canseco D, Mammen PP, Rothermel BA, Olson EN, Sadek HA (2013) Regulation of neonatal and adult mammalian heart regeneration by the miR-15 family. Proc Natl Acad Sci U S A 110:187–192

Qiao W, Zhang W, Gai Y, Zhao L, Fan J (2014) The histone acetyltransferase MOF overexpression blunts cardiac hypertrophy by targeting ROS in mice. Biochem Biophys Res Commun 448:379–384

Romano SL, Lionetti V (2013) From cell phenotype to epigenetic mechanisms: new insights into regenerating myocardium. Can J Physiol Pharmacol 91:579–585

Sanchez-Freire V, Lee AS, Hu S, Abilez OJ, Liang P, Lan F, Huber BC, Ong SG, Hong WX, Huang M, Wu JC (2014) Effect of human donor cell source on differentiation and function of cardiac induced pluripotent stem cells. J Am Coll Cardiol 64:436–448

Schlingman DJ, Mack AH, Kamenetska M, Mochrie SG, Regan L (2014) Routes to DNA accessibility: alternative pathways for nucleosome unwinding. Biophys J 107:384–392

Shikama N, Lutz W, Kretzschmar R, Sauter N, Roth JF, Marino S, Wittwer J, Scheidweiler A, Eckner R (2003) Essential function of p300 acetyltransferase activity in heart, lung and small intestine formation. EMBO J 22:5175–5185

Stein AB, Jones TA, Herron TJ, Patel SR, Day SM, Noujaim SF, Milstein ML, Klos M, Furspan PB, Jalife J, Dressler GR (2011) Loss of H3K4 methylation destabilizes gene expression patterns and physiological functions in adult murine cardiomyocytes. J Clin Invest 121:2641–2650

Tao Z, Chen B, Tan X, Zhao Y, Wang L, Zhu T, Cao K, Yang Z, Kan YW, Su H (2011) Coexpression of VEGF and angiopoietin-1 promotes angiogenesis and cardiomyocyte proliferation reduces apoptosis in porcine myocardial infarction (MI) heart. Proc Natl Acad Sci U S A 108:2064–2069

Taylor PB, Liew CC (1976) Acetylation of nuclear proteins in the isolated perfused rat heart. Basic Res Cardiol 71:27–35

Tirziu D, Giordano FJ, Simons M (2010) Cell communications in the heart. Circulation 122:928–937

Vecellio M, Spallotta F, Nanni S, Colussi C, Cencioni C, Derlet A, Bassetti B, Tilenni M, Carena MC, Farsetti A, Sbardella G, Castellano S, Mai A, Martelli F, Pompilio G, Capogrossi MC, Rossini A, Dimmeler S, Zeiher A, Gaetano C (2014) The histone acetylase activator pentadecylidenemalonate 1b rescues proliferation and differentiation in the human cardiac mesenchymal cells of type 2 diabetic patients. Diabetes 63:2132–2147

Vrijsen KR, Sluijter JP, Schuchardt MW, van Balkom BW, Noort WA, Chamuleau SA, Doevendans PA (2010) Cardiomyocyte progenitor cell-derived exosomes stimulate migration of endothelial cells. J Cell Mol Med 14:1064–1070

Wang K, Liu F, Zhou LY, Long B, Yuan SM, Wang Y, Liu CY, Sun T, Zhang XJ, Li PF (2014) The long noncoding RNA CHRF regulates cardiac hypertrophy by targeting miR-489. Circ Res 114:1377–1388

Wang X, Huang W, Liu G, Cai W, Millard RW, Wang Y, Chang J, Peng T, Fan GC (2014) Cardiomyocytes mediate anti-angiogenesis in type 2 diabetic rats through the exosomal transfer of miR-320 into endothelial cells. J Mol Cell Cardiol 74:139–150

Wang Y, Dasso M (2009) SUMOylation and deSUMOylation at a glance. J Cell Sci 122:424952

Xiao D, Dasgupta C, Chen M, Zhang K, Buchholz J, Xu Z, Zhang L (2014) Inhibition of DNA methylation reverses norepinephrine-induced cardiac hypertrophy in rats. Cardiovasc Res 101:373–382

Yang AS, Estécio MR, Doshi K, Kondo Y, Tajara EH, Issa JP (2004) A simple method for estimating global DNA methylation using bisulfite PCR of repetitive DNA elements. Nucleic Acids Res 32, e38

Zhang L, Qin X, Zhao Y, Fast L, Zhuang S, Liu P, Cheng G, Zhao TC (2012) Inhibition of histone deacetylases preserves myocardial performance and prevents cardiac remodeling through stimulation of endogenous angiomyogenesis. J Pharmacol Exp Ther 341:285–293

Zhang Y (2003) Transcriptional regulation by histone ubiquitination and deubiquitination. Genes Dev 17:2733–2740

Zhao L, Borikova AL, Ben-Yair R, Guner-Ataman B, MacRae CA, Lee RT, Burns CG, Burns CE (2014) Notch signaling regulates cardiomyocyte proliferation during zebrafish heart regeneration. Proc Natl Acad Sci U S A 111:1403–1408

Zile MR, Mehurg SM, Arroyo JE, Stroud RE, DeSantis SM, Spinale FG (2011) Relationship between the temporal profile of plasma microRNA and left ventricular remodeling in patients after myocardial infarction. Circ Cardiovasc Genet 4:614–619

Chapter 8
Stem Cell Secretome and Paracrine Activity

Felix B. Engel

8.1 Stem Cell-Based Therapy

During mammalian development the heart grows due to proliferation of cardiomyocytes. However, shortly after birth, cardiomyocytes stop to proliferate for an unknown reason (van Amerongen and Engel 2008). As newts were known to be able to regenerate their heart after injury by inducing cardiomyocyte proliferation (Oberpriller and Oberpriller 1974) and the observation of cell cycle activity in adult cardiomyocytes in diseased human hearts (van Amerongen and Engel 2008), the main focus in the field of cardiac regeneration laid on the attempt to induce proliferation of mammalian cardiomyocytes. However, there was only little success in inducing adult mammalian cardiomyocyte proliferation (Zebrowski and Engel 2013). The report by Orlic and coworkers in 2001, describing that transplantation of bone marrow cells (BMS) in infarcted mice resulted after 9 days in "newly formed myocardium occupying 68 % of the infarcted portion of the ventricle", electrified cardiologists worldwide (Orlic et al. 2001). Due to the enormous potential of BMS and the urgent need of new cardiac therapies the first stem cell-based clinical trials were initiated already between 2002 and 2005. However, not all studies observed beneficial effects, which were in addition at best modest. The differences were ascribed to the heterogeneity of BMS and differences in trial design, treatment methods, outcome evaluation, and cell isolation methods (Wollert and Drexler 2010). Consequently, numerous additional preclinical and clinical studies utilizing a large number of different stem cell types, such as mesenchymal stem cells (MSCs) and cardiac progenitor cells, were tested (Gerbin and Murry 2015; Wollert 2015; Ye and Yeghiazarians 2015). Yet, the results of these studies are controversial and it remains unclear why in some trials

F.B. Engel (✉)
Department of Nephropathology, Laboratory of Experimental Renal and Cardiovascular Research, Institute of Pathology, Friedrich-Alexander-Universität Erlangen-Nürnberg (FAU), Schwabachanlage 12, 91054, Erlangen, Germany
e-mail: felix.engel@uk-erlangen.de

R. Madonna (ed.), *Stem Cells and Cardiac Regeneration*, Stem Cell Biology and Regenerative Medicine, DOI 10.1007/978-3-319-25427-2_8

stem cells exhibited a positive effect on cardiac function and in others not (Nowbar et al. 2014; Rehman 2013; Rosen et al. 2014; Surder et al. 2013; Traverse et al. 2012). However, the fact that some of the clinical trials demonstrated a beneficial effect of stem cell-based therapies should be encouraging enough to identify the underlying mechanisms of this beneficial effect and to optimize the therapy.

8.2 The Paracrine Dogma

Initially, several data had suggested that differentiation of stem cells into cardiomyocytes is the major contribution to an improved cardiac function (Makino et al. 1999; Orlic et al. 2001; Toma et al. 2002). However, it has been demonstrated by several independent groups that the utilized adult stem cells (e.g. MSC and different kind of cardiac progenitor cells) do not differentiate into cardiomyocytes, or at least not to a significant amount (Keith and Bolli 2015; Leiker et al. 2008; Murry et al. 2004; van Berlo et al. 2014; Ye and Yeghiazarians 2015). First doubts on the importance of cardiac differentiation of transplanted stem cells were cast by the group of Victor Dzau. They demonstrated that MSCs overexpressing Akt protect the heart from ischemic injury despite low levels of cellular fusion and cardiac differentiation. Moreover, already 3 days after stem cell application the infarct size was decreased and cardiac function restored. It is unlikely that stem cells can differentiate in that time into fully functioning cardiomyocytes. Finally, they demonstrated that conditioned medium of their cells were sufficient to improve cardiac function. Consequently, they postulated that improved cardiac function "is achieved through protection of the ischemic myocardium by paracrine mediator(s) released in situ by the MSCs" (Gnecchi et al. 2005, 2006, 2008; Noiseux et al. 2006). This conclusion was further supported by a study of Shabbir and coworkers that demonstrated that also MSCs injected in the skeletal muscle improved heart function in a hamster heart failure model (Shabbir et al. 2009). The conclusion that stem cells, such as MSCs of different origin (bone marrow or adipose fat), human and mouse embryonic stem cells, and cancer stem cells, exert their beneficial effect through their secretome is meanwhile widely accepted (Makridakis et al. 2013), not only in regards to cardiac disease (Gnecchi et al. 2008; Li et al. 2014; Ranganath et al. 2012) but for example also central nervous system disorders (Paul and Anisimov 2013), renal injury (Bi et al. 2007), articular cartilage defects (Stoddart et al. 2015), and rheumatic diseases (Maumus et al. 2013).

8.3 Profiling the Stem Cell Secretome

The stem cell secretome consist of secreted proteins (e.g. growth factors, angiogenic factors, hormones, cytokines, extracellular matrix proteins, extracellular matrix proteases, and hormones) as well as genetic material (e.g. double-stranded DNA, mRNA, microRNA, and lncRNA), which is released via extracellular vesicles or exosomes (Ailawadi et al. 2015; Emanueli et al. 2015; Katsuda et al. 2013; Ranganath et al. 2012; Valadi et al. 2007).

8.3.1 Proteomics

Due to the great potential of the stem cell secretome, intensive efforts have been invested to identify the secreted factors and to characterize their function. The most accurate way to determine the secretome of the stem cells appears to be the characterization in vivo, where the stem cells exert their beneficial effect. However, even though it is possible to collect interstitial fluid in vivo by capillary microdialysis devices or ultrafiltration probes, it appears impossible to determine the cellular origin of these proteins (Brown et al. 2012). Thus, stem cell secretomes are currently analyzed under in vitro conditions. Here, it has to be considered that the culture condition can have significant influence on the stem cell secretome. For example, Uemura and coworkers observed that hypoxic preconditioning enhanced the beneficial effect of BMS due to upregulation of several survival factors (Uemura et al. 2006). In addition, it has been shown that the secretome can be modified by cell-to-cell interactions, exposure to small molecules, growth factors or cytokines and by genetic manipulation (Drago et al. 2013; Ranganath et al. 2012; Salgado et al. 2010). Finally, it has been observed, utilizing microarray analysis, that MSCs grown in 3D spheroids upregulate the expression of a number of genes including the anti-apoptotic protein stanniocalcin-1 and the anti-inflammatory protein TSG-6 (Bartosh et al. 2010).

A large set of methods are available for the characterization of the secretome. Classical methods include 1-D or 2-D gel based and chromatographic fractionations and protein identification using mass spectrometry. Newer and more advanced technologies include highly sensitive antibody-based techniques, protein microarrays, and quantitative mass spectrometry (Kupcova Skalnikova 2013). The first proteomic analysis of human MSC secretome was published in 2003 and, since then, over 30 more studies have been published (Skalnikova et al. 2011; Tran and Damaser 2015).

8.3.2 Exosomes

Secreted vessels play important roles in cell-to-cell communication. Vesicles of a size ranging in their diameter from 30 to 100 nm, called exosomes, have currently received the most attention mainly due to their potential diagnostic and therapeutic applications (Ailawadi et al. 2015; Emanueli et al. 2015; Katsuda et al. 2013; Ranganath et al. 2012; Valadi et al. 2007; Vlassov et al. 2012). Secreted vesicle can be collected by differential centrifugation and/or filtration through 0.1–0.22 mm filters. Subsequently, they can be pelleted by ultracentrifugation and exosomes can be enriched by floatation on density gradients (sucrose or iodixanol). Alternative methods are microfilters, gel filtration chromatography, flow field-flow fractionation, immunoisolation techniques, or ExoQuick, a polymer-based exosome precipitation kit (Kupcova Skalnikova 2013; Malik et al. 2013). As exosomes contain genetic material (e.g. double-stranded DNA, mRNA, microRNA, and lncRNA) it is important to expand proteomic approaches by microarray or sequencing tools to map the entire secretome (Ailawadi et al. 2015; Emanueli et al. 2015; Katsuda et al. 2013; Ranganath et al. 2012; Valadi et al. 2007).

8.3.3 Bioinformatics

Gene expression analysis combined with bioinformatics provides an alternative approach to proteomics to map the stem cell secretome. Stem cells can be collected based on specific or highly expressed cell surfaces markers via fluorescence-activated cell sorting or MACS® technology. After determining the gene expression profile by for example utilizing next generation sequencing or microarray technology, several software tools can be utilized to determine whether a gene is secreted or not if unknown. TargetP (Emanuelsson et al. 2007) and Protein Prowler (Boden and Hawkins 2005) predict based on N-terminal pre-sequences whether a candidate gene is secreted, is localized in the cytoplasm, or whether it is targeted to mitochondria or the nucleus. Yet, the more common strategy to identify secreted proteins is an algorithm of the following software tools:

1. SignalP: does the protein contain a signal peptide indicating classic secretion (Bendtsen et al. 2004b).
2. If not: SecretomeP: is it a non-classically secreted protein (Bendtsen et al. 2004a).
3. If SignalP or SecretomeP indicates the candidate protein is a secreted protein one should ensure that none of the following software tools indicate a cellular localization:
   (a) Phobius (Kall et al. 2004), TMHMM Server (Kahsay et al. 2005; Krogh et al. 2001): membrane-bound
   (b) TargetP (Emanuelsson et al. 2007), Protein Prowler (Boden and Hawkins 2005), iPSORT (Bannai et al. 2002), NetNES (la Cour et al. 2004): mitochondria/nucleus/endoplasmic reticulum reticulum

However, it is important to be aware that approximately 1/3 of proteins identified in unbiased proteomic studies lack the sequence-specific signature which is required to predict that a protein is a classical or non-classical secreted factor. This might be due to an accuracy of the software tools of <100 %. For example, currently available software tools cannot detect proteins secreted by exosomes. On the other hand, proteomic samples can be contaminated by proteins from dying or injured cells (Brown et al. 2012).

8.4 Secretomes in Cardiac Therapy

Shortly after Gnecchi and coworkers demonstrated that the beneficial effect of stem cell-based therapies is predominantly exhibited by secreted factors, Korf-Klingebiel and coworkers reported the first analysis of a secretome of stem cells used for cardiac therapy (Korf-Klingebiel et al. 2008). They isolated BMS from 15 patients with acute myocardial infarction (AMI), as previously done in the BOOST trial (Wollert et al. 2004), cultured the cells in serum-free medium, and analyzed the secretome by ProteinChip (174 probes) and GeneChip array (54,675 transcripts).

As a control they used the secretome of peripheral blood leucocytes. Their analysis revealed that based on the ProteinChip array 25 factors were up and 10 factors downregulated in BMS, while the GeneChip array data indicated that 125 secreted factors were up and 70 downregulated. The authors concluded that BMS secrete high amounts of pro-angiogenic and cytoprotective factors. In addition they suggested that "characterization of the BMC secretome may lead to an identification of factors with therapeutic potential after AMI". Nguyen and coworkers showed that the MSC secretome improved cardiac function also in a swine model of MI and confirmed utilizing protein array technology that the secretome contains various pro-angiogenic and anti-apoptotic factors. In addition, they found anti-remodeling properties (Nguyen et al. 2010).

Recently, several groups have performed detailed analyses of the stem cell secretome in order to identify novel secreted factors with beneficial effects in the heart. Liu and coworkers discovered based on an antibody array that cardiac progenitor cells abundantly secrete soluble junctional adhesion molecule-A (JAM-A). In vitro and in vivo experiments modulating JAM-A expression (siRNA, anti–JAM-A neutralizing antibody) revealed that JAM-A secreted by cardiac progenitor cells injected in the infarcted heart improves cardiac function by reducing neutrophil infiltration and preventing cardiac remodeling (Liu et al. 2014). Huang and coworkers identified C3orf58, utilizing microarray technology followed by a bioinformatics analysis, "as a novel hypoxia and Akt-induced stem cell factor (HASF)" with anti-apoptotic properties (Huang et al. 2014). Finally, Korf-Klingebiel and coworkers determined the secretome of bone marrow cells (CXCR4high, enriched by MACS® technology) from patients with acute MI from the BOOST-2 trial to identify novel secreted proteins with therapeutic potential (Korf-Klingebiel et al. 2015). Using DNA microarrays covering more than 47,000 transcripts they selected 4000 highly expressed transcripts that were poorly characterized. Based on a detailed bioinformatic secretome analysis (see also 9.3) they identified 42 putative secreted proteins. To obtain the secreted proteins the authors overexpressed the corresponding genes in HEK-293 cells and assayed for a cytoprotective effect of the conditioned supernatants using neonatal rat ventricular cardiomyocytes. This screen revealed a secreted protein encoded by an open reading frame on chromosome 19 (C19orf10) that promotes cardiomyocyte survival and angiogenesis. The authors named it myeloid-derived growth factor (MYDGF) as it is endogenously expressed by bone marrow-derived monocytes and macrophages. The authors could demonstrate utilizing MYDGF-deficient and recombinant MYDGF that MYDGF reduces scar size and contractile dysfunction after MI.

8.5 Effect of Stem Cell-Secreted Factors

The high complexity of the secretome has considerable potential for the treatment of a large variety of diseases, including heart disease, as the combination of the large number of factors can enhance or inhibit several different processes, such as apoptosis, angiogenesis, inflammation, fibrosis, remodeling, metabolic changes, cardiomyocyte proliferation, and activation/recruitment of endogenous stem cells, to prevent, reduce, and reverse cardiac injury (Fig. 8.1).

8.5.1 Angiogenesis and Revascularization

Angiogenesis is widely regarded as an attractive approach to treat ischemic heart disease, for example to meet demands for blood supply during pathological hypertrophy, peripheral artery disease, and after MI (Khurana et al. 2005; Oka et al. 2014). In recent years it has been demonstrated that pro-angiogenic factors are readily secreted by stem cells. For MSCs it has been demonstrated that it contains vascular endothelial growth factor (VEGF), fibroblast growth factor 2 (FGF-2), interleukin-6 (IL-6), placental growth factor, angiopoietin-1 (Ang-1), monocyte chemoattractant protein-1, placental growth factor (PLGF), hepatocyte growth factor (HGF), transforming growth factor β (TGF-β), MYDGF, and cysteine-rich angiogenic inducer 61 (Cyr61) (Boomsma and Geenen 2012; Chen et al. 2008; Estrada et al. 2009; Hung et al. 2007; Kinnaird et al. 2004; Korf-Klingebiel et al. 2015; Rehman et al. 2004; Wu et al. 2007). The secretion of angiogenic factors by MSCs can be upregulated by a range of stimuli such as transforming growth factor α (TGF-α) (De Luca et al. 2011), tumor necrosis factor α (TNF-α), LPS, hypoxia (Crisostomo et al. 2008; Rehman et al. 2004), epidermal growth factor (EGF) (Kilroy et al. 2007), and serum starvation (Oskowitz et al. 2011). Besides growth factors and cytokines it has been suggested stem cells release exosomes that can promote angiogenesis via microRNAs such as miR-126, miR-130a, miR-132, and miR-146 (Ailawadi et al. 2015; Cervio et al. 2015; Das and Halushka 2015; Emanueli et al. 2015). The pro-angiogenetic effect of the stem cell secretome or individual factors can be tested in assays such as endothelial cell proliferation and migration assays, HUVEC-tube-formation assay, chicken chorioallantoic membrane (CAM) assay, and the mouse matrigel plug assay (Bronckaers et al. 2014). That MSCs and their secretome promote angiogenesis in vivo after cardiac injury has been demonstrated in a large number of independent studies (e.g. Nagaya et al. 2004; Nagaya et al. 2005; Takahashi et al. 2006; Timmers et al. 2011).

8.5.2 Anti-apoptotic Effects

Ischemic diseases induce apoptosis of functional cells, such as cardiomyocytes, due to the lack of oxygen and nutrients. Thus, anti-apoptotic therapies can delay cell death providing more time for angiogenesis to supply the injured tissue. A large number of studies has shown that anti-apoptotic therapies significantly protect the heart from ischemic injury (Lavu et al. 2011). That stem cells secret in addition to pro-angiogenic also anti-apoptotic factors has been shown for MSCs by Gnecchi and coworkers

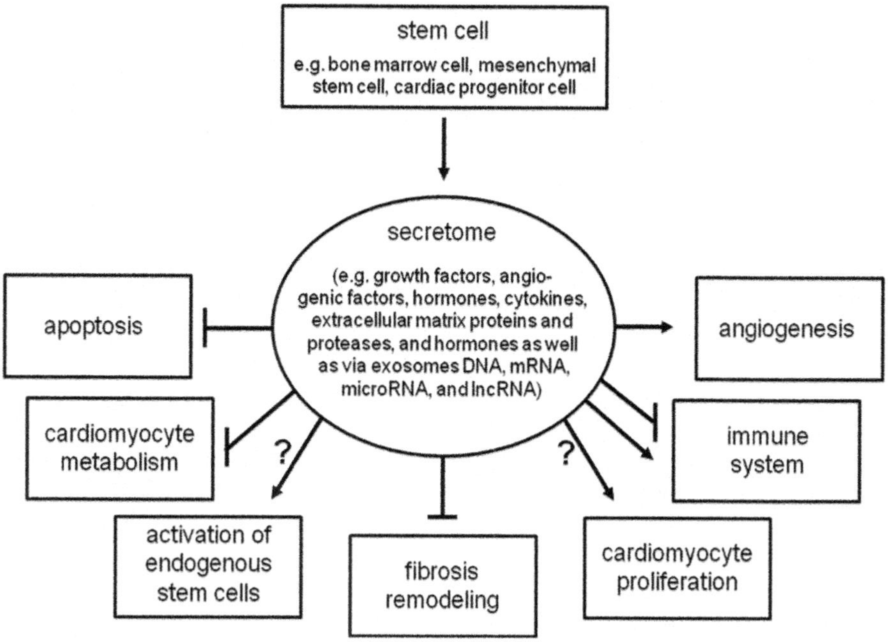

Fig. 8.1 Paracrine effect of stem cells thought to improve cardiac function after injury

(Gnecchi et al. 2005, 2009) and for adipose-derived stem cells by Rehman and coworkers (Rehman et al. 2004). Key pro-survival factors secreted by MSCs are for example the secreted frizzled related protein 2 (Sfrp2) (Mirotsou et al. 2007), insulin growth factor 1 (IGF-1), and platelet-derived growth factor PDGF (Takahashi et al. 2006). As for angiogenic factors, the cardioprotective effect of the stem cell secretome could be enhanced by a range of stimuli such with TGF-β or by activating the tumor necrosis factor receptor-2 (Herrmann et al. 2010). Finally, it has been shown that exosomes from human cardiac progenitor cells inhibit cardiomyocyte apoptosis and improve cardiac function after MI via miR-210 (Barile et al. 2014).

8.5.3 Anti-remodeling

Inflammatory cells infiltrate the heart after an MI, interstitial cells proliferate, and a subset of fibroblasts differentiates into myofibroblasts depositing extracellular matrix (ECM) resulting in the replacement of the necrotic myocardium with scar tissue. The initial inflammatory phase of scarring helps to maintain the tensile strength and integrity of the heart. Consequently, inhibition of this process leads to increased incidence of cardiac rupture or ventricular dilatation (Seropian et al. 2014). However, it has to be noted that ECM remodeling is a highly delicate and well-orchestrated process. For example, it is initially important to degrade the existing ECM by activation of latent matrix metalloproteinases (MMPs) to promote the infiltration of inflammatory cells. During scar maturation, fibroblasts and vascular cells in the scar undergo apoptosis and a collagen-rich scar is formed that promotes both infarct expansion and extension leading to overall remodeling and change in the shape of the ventricle (Leask 2015; Sun et al. 2002). Successful wound healing and remodeling after a MI thus requires a delicate balance of collagen production as well as degradation.

Even though many stem cell-based studies report reduced fibrosis and improved remodeling, it is poorly understood which factors are responsible for this improvement and what the underlying mechanism is. One major focus in modulating (e.g. after MI) or preventing (e.g. hypertension and diabetes mellitus) cardiac fibrosis is to understand the underlying mechanisms of myofibroblast differentiation. It is known that this process is highly regulated by extracellular factors such as TGF-β, endothelin-1 (ET-1), angiotensin II (Ang II), connective tissue growth factor (CCN2), and PDGF (Leask 2015). Thus, it is likely that the secretome of stem cells has also an influence on cardiac remodeling by acting on myofibroblast differentiation and thus the regulation of collagen biosynthesis and maturation. Mias and co-workers have shown that conditioned medium of MSCs reduce the proliferation of fibroblasts, represses alpha-smooth muscle actin expression and collagen secretion, and promotes matrix metalloproteinase secretion. The work of He and coworkers has shown that Sfrp2, known to be secreted by MSCs (Mirotsou et al. 2007), reduces fibrosis and improves cardiac function via its inhibition of bone morphogenetic protein 1 (Bmp1)/tolloid-like metalloproteinase, a key regulator of collagen biosynthesis and maturation after tissue injury (He et al. 2010).

8.5.4 Metabolism

The heart requires ATP for contraction and has a very high-energy demand. During heart development cardiac metabolism switches from predominantly glycolysis and glucose oxidation in fetal cardiomyocytes to β-oxidation of fatty acids in adult cardiomyocytes (Gnecchi et al. 2008; Stanley et al. 2005). Yet, after MI the heart undergoes again metabolic alterations switching back to increased glucose uptake. This results in a high-energy phosphate content and phosphocreatine-to-ATP ratio in the infarct border zone, which is believed to correspond to the severity of left ventricular contractile dysfunction and left ventricular remodeling (Hu et al. 2006; Stanley et al. 2005). Gnecchi and coworkers have shown that MSCs overexpressing Akt, but not unmodified MSCs, can prevent cardiac metabolic remodeling following MI (Gnecchi et al. 2009). Subsequently, Hughey and coworkers have reported that also transplantation of non-modified MSCs enhanced the energetic state in the infarcted heart. Utilizing hyperinsulinemic-euglycemic clamps, coupled with 2-[(14) C]deoxyglucose administration, they demonstrated that the "improved systolic performance in MSC-treated mice was associated with a preservation of in vivo insulin-stimulated cardiac glucose uptake" and insulin signaling (Hughey et al. 2014).

8.5.5 *Immune and Inflammatory Modulation*

Efforts to elucidate the underlying mechanisms of tissue regeneration revealed the importance of immune cell types in promoting a permissive local environment for effective cell replacement and restoration of tissue integrity (Forbes and Rosenthal 2014). In addition, it has been shown that modulation of the immune system can prevent cardiac remodeling. For example, IL-1 or MCP-1 knockout mice exhibit markedly reduced dilative remodeling after MI, without affecting infarct size (Bujak et al. 2008; Dewald et al. 2005). In contrast, interleukin-1 receptor-associated kinase (IRAK)-M knockout mice had enhanced adverse remodeling and worse systolic dysfunction indicating that "IRAK-M attenuates adverse postinfarction remodeling suppressing leukocyte inflammatory activity, while inhibiting fibroblast-mediated matrix degradation" (Chen et al. 2012).

Xenotransplantation experiments suggested in 2000 that MSCs possess the ability to modulate the immune system (Liechty et al. 2000). Subsequently, Di Nicola and coworkers showed that MSCs can suppress the cell-mediated immune responses by inhibiting proliferation of CD4+ and CD8+ T cells (Di Nicola et al. 2002). Meanwhile it is clear that MSCs modulate different stages of the immune system namely antigen recognition and presentation, T cell activation, proliferation, and differentiation, and the T-cell effector stage (Aurora and Olson 2014; Kim and Cho 2015; Liang et al. 2014). For example, by releasing cytokines such as HLA-G5, prostaglandin E2 (PGE2), and indoleamine 2,3-dioxygenase (IDO) MSCs modulate the differentiation of monocytes. Secretion of HGF, IDO, PGE2, inducible nitric-oxide synthase, TGFβ1, and heme oxygenase-1 blocks T-cell activity (Mandi and Vecsei 2012; Uccelli et al. 2008; Zhao et al. 2008). In the beginning it was believed that MSCs are solely immunosuppressive (Fibbe et al. 2007; Selmani et al. 2008). For example, Lee and coworkers showed that intravenous human MSCs improve MI in mice because MSCs embolized in the lung are activated to secrete the powerful anti-inflammatory factor tumor necrosis factor-stimulated gene 6 protein (TSG-6) (Lee et al. 2009). Yet, it has become clear that they can also stimulate the release of pro-inflammatory molecules (Bernardo and Fibbe 2013; Frangogiannis 2012).

8.5.6 *Activation of Endogenous Repair Processes*

The heart has for a long time been considered as a post-mitotic organ, meaning that shortly after birth no new cardiomyocytes are formed, neither by cardiomyocyte proliferation nor by stem cell differentiation. However, recent data suggest that there is modest cardiomyocyte turnover (around 1 %) in adult mouse and human hearts (Bergmann et al. 2015; Senyo et al. 2013). The current hypothesis is that new cardiomyocytes are predominantly added by proliferation of pre-existing cardiomyocytes. Recently, fate mapping of hypoxic cells and their progenies identified a "rare population of hypoxic cardiomyocytes that display characteristics of proliferative neonatal cardiomyocytes, such as smaller size, mononucleation and lower oxidative DNA damage" (Kimura et al. 2015). These cells might present a subpopulation of cardiomyocytes utilized for cardiac homeostasis. Whether the secretome can promote endogenous stem cells to differentiate into cardiomyocytes or enhances proliferation of a subpopulation of cardiomyocytes remains unclear. However, it has for example been shown that factors such as FGF1 can promote cardiomyocyte proliferation (Engel et al. 2005) and improve cardiac function (Engel et al. 2006). Even though some studies indicate cell cycle activation of cardiomyocytes upon stem cell transplantation there is currently no evidence for the formation of new cardiomyocytes in stem cell-based therapies.

8.6 Cell-Free Therapy

Growth factors and cytokines have due to their pleiotropic effects long before stem-cell based approaches been considered for heart therapies. For example, the beta isoform of Neuregulin-1 (NRG-1β) exhibits anti-fibrotic effects in several animal models of heart failure (Galindo et al. 2014b) including diabetic cardiomyopathy in a rat model (Li et al. 2013) and MI in swine (Galindo et al. 2014a). NRG-1β has also been demonstrated in clinical trials to have beneficial effects by maintaining cardiac structure and function, as well as mediating reverse remodeling (Galindo et al. 2014b; Rupert and Coulombe 2015). NRG-1β exhibits its positive effects for example by enhancing cardiomyocyte cell survival (Fukazawa et al. 2003), angiogenesis (Russell et al. 1999), and excitation-coupling (Timolati et al. 2006). In addition, Galindo and coworkers have demonstrated that NRG-1β reduces the number of myofibroblasts (Galindo et al. 2014a; Pentassuglia and Sawyer 2013).

Several clinical trials have been initiated to investigate the potential of cytokines and growth factors such as VEGF-165, VEGF-121, VEGF-2 (VEGF-C), FGF-4, granulocyte colony stimulating factor (GCSF), EPO, and IGF-1 for treating cardiovascular diseases focusing on angiogenesis (Beohar et al. 2010; Engelmann et al. 2006; Formiga et al. 2012; Hwang and Kloner 2010; Ripa et al. 2006). However, the results have been disappointing with no or minor effects, which has been attributed, at least partially, to the short-lived effect and high instability of proteins after intravenous infusion or intracoronary delivery (Formiga et al. 2012). A major problem in these therapies besides limited protein stability is the need of a distinct dose at a distinct location of a factor to exert a positive effect. For example, pro-angiogenic factors in to high doses can lead to the formation of aberrant and leaky vessels (Carmeliet 2005), hypotension (Eppler et al. 2002), and tumor angiogenesis (Epstein et al. 2001). In addition, the timing of therapy is an important factor. For example, IL6, a pleiotropic cytokine activating immune cells, acts after an MI protective and limits cardiac damage. Yet, if IL6 remains chronically activated it is pathogenic leading to chronic inflammation and fibrosis (Fontes et al. 2015). The importance of the positive early effect has been demonstrated by catastrophic effects upon broad inhibition of the inflammatory reaction in patients with MI using systemic corticosteroids (Roberts et al. 1976). Subsequent studies targeting specific immune pathways demonstrated that inhibition of adhesion molecules, chemokines and cytokines protects the ischemic and reperfused myocardium, reducing the size of the infarct by 40–50 %. Yet, clinical trials did not meet the expectations (Frangogiannis 2014). Another good example for the importance of time point of injection and duration of a treatment has been provided by Kang and coworkers describing the efficiency of GCSF therapy depending on timing and duration of the therapy (Kang et al. 2004).

Collectively, it is important to optimize the delivery strategy of secreted factors in regards to concentration and location. Recently first strategies have been developed. For example, gene constructs have been designed that allow cardiac-specific expression upon hypoxia (Su et al. 2004; Tang 2005). Such constructs can be delivered by adeno-associated viruses utilizing a catheter or as plasmids bound to microbubbles, which can be released by ultrasound-targeted microbubble destruction (Fujii et al. 2009; Suzuki et al. 2008). In addition, novel materials are continuously developed for controlled release of proteins and DNA (Jo and Tabata 2015; Yockman et al. 2008).

8.7 Outlook

Taken together, neither the delivery of single growth factors or cytokines nor the use of stem cells such as MSCs have been shown to have a significant beneficial effect on cardiac function. However, stem cells can supply secreted factors with the advantage of sustained pharmacokinetics, synergy from multiple factors, and opportunity for systemic infusion and thus repeated dosing. Thus, a valid approach to treat cardiovascular diseases might be the enhancement of stem cells by preconditioning or genetic manipulation to design cells that secrete multiple cytokines/growth factors at different concentrations to act synergistically at a distinct localization (MSCs will be homed to the correct place). These cells are called "Second-generation" stem cells (Nunez Garcia et al. 2015; Tran and Damaser 2015) and could be thought of as combinatorial drug manufacture and delivery mechanism. In addition, the analysis of stem cell secretomes holds promise to identify additional important factors for cardiac repair.

However, stem cell-based therapies have significant disadvantages related to immune compatibility, tumorigenicity, and the transmission of infections and they are time consuming (stem cell expansion and maintenance). Thus, it is recommended to continue to work on multi-factor therapies focusing on establishment of new technologies to overcome hurdles like protein stability, local application, release kinetics, and concentration. This would allow the preparation of large quantities of doses of cytokine/growth factor cocktails in an allogeneic or off-the-shelf fashion and thus treatment when needed including in acute conditions such as MI.

References

Ailawadi S, Wang X, Gu H, Fan GC (2015) Pathologic function and therapeutic potential of exosomes in cardiovascular disease. Biochim Biophys Acta 1852:1–11
Aurora AB, Olson EN (2014) Immune modulation of stem cells and regeneration. Cell Stem Cell 15:14–25
Bannai H, Tamada Y, Maruyama O, Nakai K, Miyano S (2002) Extensive feature detection of N-terminal protein sorting signals. Bioinformatics 18:298–305
Barile L, Lionetti V, Cervio E, Matteucci M, Gherghiceanu M, Popescu LM, Torre T, Siclari F, Moccetti T, Vassalli G (2014) Extracellular vesicles from human cardiac progenitor cells inhibit cardiomyocyte apoptosis and improve cardiac function after myocardial infarction. Cardiovasc Res 103:530–541
Bartosh TJ, Ylostalo JH, Mohammadipoor A, Bazhanov N, Coble K, Claypool K, Lee RH, Choi H, Prockop DJ (2010) Aggregation of human mesenchymal stromal cells (MSCs) into 3D spheroids enhances their antiinflammatory properties. Proc Natl Acad Sci U S A 107:13724–13729
Bendtsen JD, Jensen LJ, Blom N, Von Heijne G, Brunak S (2004a) Feature-based prediction of non-classical and leaderless protein secretion. Protein Eng Des Sel 17:349–356
Bendtsen JD, Nielsen H, von Heijne G, Brunak S (2004b) Improved prediction of signal peptides: SignalP 3.0. J Mol Biol 340:783–795
Beohar N, Rapp J, Pandya S, Losordo DW (2010) Rebuilding the damaged heart: the potential of cytokines and growth factors in the treatment of ischemic heart disease. J Am Coll Cardiol 56:1287–1297
Bergmann O, Zdunek S, Felker A, Salehpour M, Alkass K, Bernard S, Sjostrom SL, Szewczykowska M, Jackowska T, Dos Remedios C, Malm T, Andrä M, Jashari R, Nyengaard JR, Possnert G, Jovinge S, Druid H, Frisén J (2015) Dynamics of cell generation and turnover in the human heart. Cell 161:1566–1575
Bernardo ME, Fibbe WE (2013) Mesenchymal stromal cells: sensors and switchers of inflammation. Cell Stem Cell 13:392–402
Bi B, Schmitt R, Israilova M, Nishio H, Cantley LG (2007) Stromal cells protect against acute tubular injury via an endocrine effect. J Am Soc Nephrol 18:2486–2496
Boden M, Hawkins J (2005) Prediction of subcellular localization using sequence-biased recurrent networks. Bioinformatics 21:2279–2286
Boomsma RA, Geenen DL (2012) Mesenchymal stem cells secrete multiple cytokines that promote angiogenesis and have contrasting effects on chemotaxis and apoptosis. PLoS One 7, e35685
Bronckaers A, Hilkens P, Martens W, Gervois P, Ratajczak J, Struys T, Lambrichts I (2014) Mesenchymal stem/stromal cells as a pharmacological and therapeutic approach to accelerate angiogenesis. Pharmacol Ther 143:181–196
Brown KJ, Formolo CA, Seol H, Marathi RL, Duguez S, An E, Pillai D, Nazarian J, Rood BR, Hathout Y (2012) Advances in the proteomic investigation of the cell secretome. Expert Rev Proteomics 9:337–345
Bujak M, Dobaczewski M, Chatila K, Mendoza LH, Li N, Reddy A, Frangogiannis NG (2008) Interleukin-1 receptor type I signaling critically regulates infarct healing and cardiac remodeling. Am J Pathol 173:57–67
Carmeliet P (2005) VEGF as a key mediator of angiogenesis in cancer. Oncology 69:4–10
Cervio E, Barile L, Moccetti T, Vassalli G (2015) Exosomes for intramyocardial intercellular communication. Stem Cells Int 2015:482171
Chen X, Lin X, Zhao J, Shi W, Zhang H, Wang Y, Kan B, Du L, Wang B, Wei Y, Liu Y, Zhao X (2008) A tumor-selective biotherapy with prolonged impact on established metastases based on cytokine gene-engineered MSCs. Mol Ther 16:749–756
Chen W, Saxena A, Li N, Sun J, Gupta A, Lee DW, Tian Q, Dobaczewski M, Frangogiannis NG (2012) Endogenous IRAK-M attenuates postinfarction remodeling through effects on macrophages and fibroblasts. Arterioscler Thromb Vasc Biol 32:2598–2608

Crisostomo PR, Wang Y, Markel TA, Wang M, Lahm T, Meldrum DR (2008) Human mesenchymal stem cells stimulated by TNF-alpha, LPS, or hypoxia produce growth factors by an NF kappa B- but not JNK-dependent mechanism. Am J Physiol Cell Physiol 294:C675–C682

Das S, Halushka MK (2015) Extracellular vesicle microRNA transfer in cardiovascular disease. Cardiovasc Pathol 24:199–206

De Luca A, Gallo M, Aldinucci D, Ribatti D, Lamura L, D'Alessio A, De Filippi R, Pinto A, Normanno N (2011) Role of the EGFR ligand/receptor system in the secretion of angiogenic factors in mesenchymal stem cells. J Cell Physiol 226:2131–2138

Dewald O, Zymek P, Winkelmann K, Koerting A, Ren G, Abou-Khamis T, Michael LH, Rollins BJ, Entman ML, Frangogiannis NG (2005) CCL2/Monocyte Chemoattractant Protein-1 regulates inflammatory responses critical to healing myocardial infarcts. Circ Res 96:881–889

Di Nicola M, Carlo-Stella C, Magni M, Milanesi M, Longoni PD, Matteucci P, Grisanti S, Gianni AM (2002) Human bone marrow stromal cells suppress T-lymphocyte proliferation induced by cellular or nonspecific mitogenic stimuli. Blood 99:3838–3843

Drago D, Cossetti C, Iraci N, Gaude E, Musco G, Bachi A, Pluchino S (2013) The stem cell secretome and its role in brain repair. Biochimie 95:2271–2285

Emanueli C, Shearn AI, Angelini GD, Sahoo S (2015) Exosomes and exosomal miRNAs in cardiovascular protection and repair. Vascul Pharmacol 71:24–30

Emanuelsson O, Brunak S, von Heijne G, Nielsen H (2007) Locating proteins in the cell using TargetP, SignalP and related tools. Nat Protoc 2:953–971

Engel FB, Schebesta M, Duong MT, Lu G, Ren S, Madwed JB, Jiang H, Wang Y, Keating MT (2005) p38 MAP kinase inhibition enables proliferation of adult mammalian cardiomyocytes. Genes Dev 19:1175–1187

Engel FB, Hsieh PC, Lee RT, Keating MT (2006) FGF1/p38 MAP kinase inhibitor therapy induces cardiomyocyte mitosis, reduces scarring, and rescues function after myocardial infarction. Proc Natl Acad Sci U S A 103:15546–15551

Engelmann MG, Theiss HD, Hennig-Theiss C, Huber A, Wintersperger BJ, Werle-Ruedinger AE, Schoenberg SO, Steinbeck G, Franz WM (2006) Autologous bone marrow stem cell mobilization induced by granulocyte colony-stimulating factor after subacute ST-segment elevation myocardial infarction undergoing late revascularization: final results from the G-CSF-STEMI (Granulocyte Colony-Stimulating Factor ST-Segment Elevation Myocardial Infarction) trial. J Am Coll Cardiol 48:1712–1721

Eppler SM, Combs DL, Henry TD, Lopez JJ, Ellis SG, Yi JH, Annex BH, McCluskey ER, Zioncheck TF (2002) A target-mediated model to describe the pharmacokinetics and hemodynamic effects of recombinant human vascular endothelial growth factor in humans. Clin Pharmacol Ther 72:20–32

Epstein SE, Kornowski R, Fuchs S, Dvorak HF (2001) Angiogenesis therapy: amidst the hype, the neglected potential for serious side effects. Circulation 104:115–119

Estrada R, Li N, Sarojini H, An J, Lee MJ, Wang E (2009) Secretome from mesenchymal stem cells induces angiogenesis via Cyr61. J Cell Physiol 219:563–571

Fibbe WE, Nauta AJ, Roelofs H (2007) Modulation of immune responses by mesenchymal stem cells. Ann N Y Acad Sci 1106:272–278

Fontes JA, Rose NR, Cihakova D (2015) The varying faces of IL-6: from cardiac protection to cardiac failure. Cytokine 74:62–68

Forbes SJ, Rosenthal N (2014) Preparing the ground for tissue regeneration: from mechanism to therapy. Nat Med 20:857–869

Formiga FR, Tamayo E, Simon-Yarza T, Pelacho B, Prosper F, Blanco-Prieto MJ (2012) Angiogenic therapy for cardiac repair based on protein delivery systems. Heart Fail Rev 17:449–473

Frangogiannis NG (2012) Regulation of the inflammatory response in cardiac repair. Circ Res 110:159–173

Frangogiannis NG (2014) The immune system and the remodeling infarcted heart: cell biological insights and therapeutic opportunities. J Cardiovasc Pharmacol 63:185–195

Fujii H, Sun Z, Li SH, Wu J, Fazel S, Weisel RD, Rakowski H, Lindner J, Li RK (2009) Ultrasound-targeted gene delivery induces angiogenesis after a myocardial infarction in mice. JACC Cardiovasc Imaging 2:869–879

Fukazawa R, Miller TA, Kuramochi Y, Frantz S, Kim YD, Marchionni MA, Kelly RA, Sawyer DB (2003) Neuregulin-1 protects ventricular myocytes from anthracycline-induced apoptosis via erbB4-dependent activation of PI3-kinase/Akt. J Mol Cell Cardiol 35:1473–1479

Galindo CL, Kasasbeh E, Murphy A, Ryzhov S, Lenihan S, Ahmad FA, Williams P, Nunnally A, Adcock J, Song Y, Harrell FE, Tran TL, Parry TJ, Iaci J, Ganguly A, Feoktistov I, Stephenson MK, Caggiano AO, Sawyer DB, Cleator JH (2014a) Anti-remodeling and anti-fibrotic effects of the neuregulin-1beta glial growth factor 2 in a large animal model of heart failure. J Am Heart Assoc 3, e000773

Galindo CL, Ryzhov S, Sawyer DB (2014b) Neuregulin as a heart failure therapy and mediator of reverse remodeling. Curr Heart Fail Rep 11:40–49

Gerbin KA, Murry CE (2015) The winding road to regenerating the human heart. Cardiovasc Pathol 24:133–140

Gnecchi M, He H, Liang OD, Melo LG, Morello F, Mu H, Noiseux N, Zhang L, Pratt RE, Ingwall JS, Ingwall JS, Dzau VJ (2005) Paracrine action accounts for marked protection of ischemic heart by Akt-modified mesenchymal stem cells. Nat Med 11:367–368

Gnecchi M, He H, Noiseux N, Liang OD, Zhang L, Morello F, Mu H, Melo LG, Pratt RE, Ingwall JS, Dzau VJ (2006) Evidence supporting paracrine hypothesis for Akt-modified mesenchymal stem cell-mediated cardiac protection and functional improvement. FASEB J 20:661–669

Gnecchi M, Zhang Z, Ni A, Dzau VJ (2008) Paracrine mechanisms in adult stem cell signaling and therapy. Circ Res 103:1204–1219

Gnecchi M, He H, Melo LG, Noiseaux N, Morello F, de Boer RA, Zhang L, Pratt RE, Dzau VJ, Ingwall JS (2009) Early beneficial effects of bone marrow-derived mesenchymal stem cells overexpressing Akt on cardiac metabolism after myocardial infarction. Stem Cells 27:971–979

He W, Zhang L, Ni A, Zhang Z, Mirotsou M, Mao L, Pratt RE, Dzau VJ (2010) Exogenously administered secreted frizzled related protein 2 (Sfrp2) reduces fibrosis and improves cardiac function in a rat model of myocardial infarction. Proc Natl Acad Sci U S A 107:21110–21115

Herrmann JL, Wang Y, Abarbanell AM, Weil BR, Tan J, Meldrum DR (2010) Preconditioning mesenchymal stem cells with transforming growth factor-alpha improves mesenchymal stem cell-mediated cardioprotection. Shock 33:24–30

Hu Q, Wang X, Lee J, Mansoor A, Liu J, Zeng L, Swingen C, Zhang G, Feygin J, Ochiai K, Bransford TL, From AH, Bache RJ, Zhang J (2006) Profound bioenergetic abnormalities in peri-infarct myocardial regions. Am J Physiol Heart Circ Physiol 291:H648–H657

Huang J, Guo J, Beigi F, Hodgkinson CP, Facundo HT, Zhang Z, Espinoza-Derout J, Zhou X, Pratt RE, Mirotsou M, Dzau VJ (2014) HASF is a stem cell paracrine factor that activates PKC epsilon mediated cytoprotection. J Mol Cell Cardiol 66:157–164

Hughey CC, James FD, Ma L, Bracy DP, Wang Z, Wasserman DH, Rottman JN, Shearer J (2014) Diminishing impairments in glucose uptake, mitochondrial content, and ADP-stimulated oxygen flux by mesenchymal stem cell therapy in the infarcted heart. Am J Physiol Cell Physiol 306:C19–C27

Hung SC, Pochampally RR, Chen SC, Hsu SC, Prockop DJ (2007) Angiogenic effects of human multipotent stromal cell conditioned medium activate the PI3K-Akt pathway in hypoxic endothelial cells to inhibit apoptosis, increase survival, and stimulate angiogenesis. Stem Cells 25:2363–2370

Hwang H, Kloner RA (2010) Improving regenerating potential of the heart after myocardial infarction: factor-based approach. Life Sci 86:461–472

Jo JI, Tabata Y (2015) How controlled release technology can aid gene delivery. Expert Opin Drug Deliv [Epub ahead of print]

Kahsay RY, Gao G, Liao L (2005) An improved hidden Markov model for transmembrane protein detection and topology prediction and its applications to complete genomes. Bioinformatics 21:1853–1858

Kall L, Krogh A, Sonnhammer EL (2004) A combined transmembrane topology and signal peptide prediction method. J Mol Biol 338:1027–1036

Kang HJ, Kim HS, Zhang SY, Park KW, Cho HJ, Koo BK, Kim YJ, Soo Lee D, Sohn DW, Han KS, Oh BH, Lee MM, Park YB (2004) Effects of intracoronary infusion of peripheral blood stem-cells mobilised with granulocyte-colony stimulating factor on left ventricular systolic function and restenosis after coronary stenting in myocardial infarction: the MAGIC cell randomised clinical trial. Lancet 363:751–756

Katsuda T, Kosaka N, Takeshita F, Ochiya T (2013) The therapeutic potential of mesenchymal stem cell-derived extracellular vesicles. Proteomics 13:1637–1653

Keith MC, Bolli R (2015) "String theory" of c-kit(pos) cardiac cells: a new paradigm regarding the nature of these cells that may reconcile apparently discrepant results. Circ Res 116:1216–1230

Khurana R, Simons M, Martin JF, Zachary IC (2005) Role of angiogenesis in cardiovascular disease: a critical appraisal. Circulation 112:1813–1824

Kilroy GE, Foster SJ, Wu X, Ruiz J, Sherwood S, Heifetz A, Ludlow JW, Stricker DM, Potiny S, Green P, Halvorsen YD, Cheatham B, Storms RW, Gimble JM (2007) Cytokine profile of human adipose-derived stem cells: expression of angiogenic, hematopoietic, and proinflammatory factors. J Cell Physiol 212:702–709

Kim N, Cho SG (2015) New strategies for overcoming limitations of mesenchymal stem cell-based immune modulation. Int J Stem Cells 8:54–68

Kimura W, Xiao F, Canseco DC, Muralidhar S, Thet S, Zhang HM, Abdulrahman Y, Chen R, Garcia JA, Shelton JM, Richardson JA, Ashour AM, Asaithamby A, Liang H, Xing C, Lu Z, Zhang CC, Sadek HA (2015) Hypoxia fate mapping identifies cycling cardiomyocytes in the adult heart. Nature 523:226–230

Kinnaird T, Stabile E, Burnett MS, Shou M, Lee CW, Barr S, Fuchs S, Epstein SE (2004) Local delivery of marrow-derived stromal cells augments collateral perfusion through paracrine mechanisms. Circulation 109:1543–1549

Korf-Klingebiel M, Kempf T, Sauer T, Brinkmann E, Fischer P, Meyer GP, Ganser A, Drexler H, Wollert KC (2008) Bone marrow cells are a rich source of growth factors and cytokines: implications for cell therapy trials after myocardial infarction. Eur Heart J 29:2851–2858

Korf-Klingebiel M, Reboll MR, Klede S, Brod T, Pich A, Polten F, Napp LC, Bauersachs J, Ganser A, Brinkmann E, Reimann I, Kempf T, Niessen HW, Mizrahi J, Schönfeld HJ, Iglesias A, Bobadilla M, Wang Y, Wollert KC (2015) Myeloid-derived growth factor (C19orf10) mediates cardiac repair following myocardial infarction. Nat Med 21:140–149

Krogh A, Larsson B, von Heijne G, Sonnhammer EL (2001) Predicting transmembrane protein topology with a hidden Markov model: application to complete genomes. J Mol Biol 305:567–580

Kupcova Skalnikova H (2013) Proteomic techniques for characterisation of mesenchymal stem cell secretome. Biochimie 95:2196–2211

la Cour T, Kiemer L, Molgaard A, Gupta R, Skriver K, Brunak S (2004) Analysis and prediction of leucine-rich nuclear export signals. Protein Eng Des Sel 17:527–536

Lavu M, Gundewar S, Lefer DJ (2011) Gene therapy for ischemic heart disease. J Mol Cell Cardiol 50:742–750

Leask A (2015) Getting to the heart of the matter: new insights into cardiac fibrosis. Circ Res 116:1269–1276

Lee RH, Pulin AA, Seo MJ, Kota DJ, Ylostalo J, Larson BL, Semprun-Prieto L, Delafontaine P, Prockop DJ (2009) Intravenous hMSCs improve myocardial infarction in mice because cells embolized in lung are activated to secrete the anti-inflammatory protein TSG-6. Cell Stem Cell 5:54–63

Leiker M, Suzuki G, Iyer VS, Canty JM Jr, Lee T (2008) Assessment of a nuclear affinity labeling method for tracking implanted mesenchymal stem cells. Cell Transplant 17:911–922

Li B, Xiao J, Li Y, Zhang J, Zeng M (2013) Gene transfer of human neuregulin-1 attenuates ventricular remodeling in diabetic cardiomyopathy rats. Exp Ther Med 6:1105–1112

Li N, Wang C, Jia L, Du J (2014) Heart regeneration, stem cells, and cytokines. Regen Med Res 2:6

Liang X, Ding Y, Zhang Y, Tse HF, Lian Q (2014) Paracrine mechanisms of mesenchymal stem cell-based therapy: current status and perspectives. Cell Transplant 23:1045–1059
Liechty KW, MacKenzie TC, Shaaban AF, Radu A, Moseley AM, Deans R, Marshak DR, Flake AW (2000) Human mesenchymal stem cells engraft and demonstrate site-specific differentiation after in utero transplantation in sheep. Nat Med 6:1282–1286
Liu ML, Nagai T, Tokunaga M, Iwanaga K, Matsuura K, Takahashi T, Kanda M, Kondo N, Naito AT, Komuro I, Kobayashi Y (2014) Anti-inflammatory peptides from cardiac progenitors ameliorate dysfunction after myocardial infarction. J Am Heart Assoc 3, e001101
Makino S, Fukuda K, Miyoshi S, Konishi F, Kodama H, Pan J, Sano M, Takahashi T, Hori S, Abe H, Hata J, Umezawa A, Ogawa S (1999) Cardiomyocytes can be generated from marrow stromal cells in vitro. J Clin Invest 103:697–705
Makridakis M, Roubelakis MG, Vlahou A (2013) Stem cells: insights into the secretome. Biochim Biophys Acta 1834:2380–2384
Malik ZA, Kott KS, Poe AJ, Kuo T, Chen L, Ferrara KW, Knowlton AA (2013) Cardiac myocyte exosomes: stability, HSP60, and proteomics. Am J Physiol Heart Circ Physiol 304:H954–H965
Mandi Y, Vecsei L (2012) The kynurenine system and immunoregulation. J Neural Transm 119:197–209
Maumus M, Jorgensen C, Noel D (2013) Mesenchymal stem cells in regenerative medicine applied to rheumatic diseases: role of secretome and exosomes. Biochimie 95:2229–2234
Mirotsou M, Zhang Z, Deb A, Zhang L, Gnecchi M, Noiseux N, Mu H, Pachori A, Dzau V (2007) Secreted frizzled related protein 2 (Sfrp2) is the key Akt-mesenchymal stem cell-released paracrine factor mediating myocardial survival and repair. Proc Natl Acad Sci U S A 104:1643–1648
Murry CE, Soonpaa MH, Reinecke H, Nakajima H, Nakajima HO, Rubart M, Pasumarthi KB, Virag JI, Bartelmez SH, Poppa V, Bradford G, Dowell JD, Williams DA, Field LJ (2004) Haematopoietic stem cells do not transdifferentiate into cardiac myocytes in myocardial infarcts. Nature 428:664–668
Nagaya N, Fujii T, Iwase T, Ohgushi H, Itoh T, Uematsu M, Yamagishi M, Mori H, Kangawa K, Kitamura S (2004) Intravenous administration of mesenchymal stem cells improves cardiac function in rats with acute myocardial infarction through angiogenesis and myogenesis. Am J Physiol Heart Circ Physiol 287:H2670–H2676
Nagaya N, Kangawa K, Itoh T, Iwase T, Murakami S, Miyahara Y, Fujii T, Uematsu M, Ohgushi H, Yamagishi M, Tokudome T, Mori H, Miyatake K, Kitamura S (2005) Transplantation of mesenchymal stem cells improves cardiac function in a rat model of dilated cardiomyopathy. Circulation 112:1128–1135
Nguyen BK, Maltais S, Perrault LP, Tanguay JF, Tardif JC, Stevens LM, Borie M, Harel F, Mansour S, Noiseux N (2010) Improved function and myocardial repair of infarcted heart by intracoronary injection of mesenchymal stem cell-derived growth factors. J Cardiovasc Transl Res 3:547–558
Noiseux N, Gnecchi M, Lopez-Ilasaca M, Zhang L, Solomon SD, Deb A, Dzau VJ, Pratt RE (2006) Mesenchymal stem cells overexpressing Akt dramatically repair infarcted myocardium and improve cardiac function despite infrequent cellular fusion or differentiation. Mol Ther 14:840–850
Nowbar AN, Mielewczik M, Karavassilis M, Dehbi HM, Shun-Shin MJ, Jones S, Howard JP, Cole GD, Francis DP (2014) Discrepancies in autologous bone marrow stem cell trials and enhancement of ejection fraction (DAMASCENE): weighted regression and meta-analysis. BMJ 348:g2688
Nunez Garcia A, Sanz-Ruiz R, Fernandez Santos ME, Fernandez-Aviles F (2015) "Second-generation" stem cells for cardiac repair. World J Stem Cells 7:352–367
Oberpriller JO, Oberpriller JC (1974) Response of the adult newt ventricle to injury. J Exp Zool 187:249–253
Oka T, Akazawa H, Naito AT, Komuro I (2014) Angiogenesis and cardiac hypertrophy: maintenance of cardiac function and causative roles in heart failure. Circ Res 114:565–571

Orlic D, Kajstura J, Chimenti S, Jakoniuk I, Anderson SM, Li B, Pickel J, McKay R, Nadal-Ginard B, Bodine DM, Leri A, Anversa P (2001) Bone marrow cells regenerate infarcted myocardium. Nature 410:701–705

Oskowitz A, McFerrin H, Gutschow M, Carter ML, Pochampally R (2011) Serum-deprived human multipotent mesenchymal stromal cells (MSCs) are highly angiogenic. Stem Cell Res 6:215–225

Paul G, Anisimov SV (2013) The secretome of mesenchymal stem cells: potential implications for neuroregeneration. Biochimie 95:2246–2256

Pentassuglia L, Sawyer DB (2013) ErbB/integrin signaling interactions in regulation of myocardial cell-cell and cell-matrix interactions. Biochim Biophys Acta 1833:909–916

Ranganath SH, Levy O, Inamdar MS, Karp JM (2012) Harnessing the mesenchymal stem cell secretome for the treatment of cardiovascular disease. Cell Stem Cell 10:244–258

Rehman J (2013) Bone marrow tinctures for cardiovascular disease: lost in translation. Circulation 127:1935–1937

Rehman J, Traktuev D, Li J, Merfeld-Clauss S, Temm-Grove CJ, Bovenkerk JE, Pell CL, Johnstone BH, Considine RV, March KL (2004) Secretion of angiogenic and antiapoptotic factors by human adipose stromal cells. Circulation 109:1292–1298

Ripa RS, Jorgensen E, Wang Y, Thune JJ, Nilsson JC, Sondergaard L, Johnsen HE, Kober L, Grande P, Kastrup J (2006) Stem cell mobilization induced by subcutaneous granulocyte-colony stimulating factor to improve cardiac regeneration after acute ST-elevation myocardial infarction: result of the double-blind, randomized, placebo-controlled stem cells in myocardial infarction (STEMMI) trial. Circulation 113:1983–1992

Roberts R, DeMello V, Sobel BE (1976) Deleterious effects of methylprednisolone in patients with myocardial infarction. Circulation 53:I204–I206

Rosen MR, Myerburg RJ, Francis DP, Cole GD, Marban E (2014) Translating stem cell research to cardiac disease therapies: pitfalls and prospects for improvement. J Am Coll Cardiol 64:922–937

Rupert CE, Coulombe KL (2015) The roles of neuregulin-1 in cardiac development, homeostasis, and disease. Biomark Insights 10:1–9

Russell KS, Stern DF, Polverini PJ, Bender JR (1999) Neuregulin activation of ErbB receptors in vascular endothelium leads to angiogenesis. Am J Physiol 277:H2205–H2211

Salgado AJ, Reis RL, Sousa NJ, Gimble JM (2010) Adipose tissue derived stem cells secretome: soluble factors and their roles in regenerative medicine. Curr Stem Cell Res Ther 5:103–110

Selmani Z, Naji A, Zidi I, Favier B, Gaiffe E, Obert L, Borg C, Saas P, Tiberghien P, Rouas-Freiss N, Carosella ED, Deschaseaux F (2008) Human leukocyte antigen-G5 secretion by human mesenchymal stem cells is required to suppress T lymphocyte and natural killer function and to induce CD4+CD25highFOXP3+ regulatory T cells. Stem Cells 26:212–222

Senyo SE, Steinhauser ML, Pizzimenti CL, Yang VK, Cai L, Wang M, Wu TD, Guerquin-Kern JL, Lechene CP, Lee RT (2013) Mammalian heart renewal by pre-existing cardiomyocytes. Nature 493:433–436

Seropian IM, Toldo S, Van Tassell BW, Abbate A (2014) Anti-inflammatory strategies for ventricular remodeling following ST-segment elevation acute myocardial infarction. J Am Coll Cardiol 63:1593–1603

Shabbir A, Zisa D, Suzuki G, Lee T (2009) Heart failure therapy mediated by the trophic activities of bone marrow mesenchymal stem cells: a noninvasive therapeutic regimen. Am J Physiol Heart Circ Physiol 296:H1888–H1897

Skalnikova H, Motlik J, Gadher SJ, Kovarova H (2011) Mapping of the secretome of primary isolates of mammalian cells, stem cells and derived cell lines. Proteomics 11:691–708

Stanley WC, Recchia FA, Lopaschuk GD (2005) Myocardial substrate metabolism in the normal and failing heart. Physiol Rev 85:1093–1129

Stoddart MJ, Bara J, Alini M (2015) Cells and secretome—towards endogenous cell re-activation for cartilage repair. Adv Drug Deliv Rev 84:135–145

Su H, Joho S, Huang Y, Barcena A, Arakawa-Hoyt J, Grossman W, Kan YW (2004) Adeno-associated viral vector delivers cardiac-specific and hypoxia-inducible VEGF expression in ischemic mouse hearts. Proc Natl Acad Sci U S A 101:16280–16285
Sun Y, Kiani MF, Postlethwaite AE, Weber KT (2002) Infarct scar as living tissue. Basic Res Cardiol 97:343–347
Surder D, Manka R, Lo Cicero V, Moccetti T, Rufibach K, Soncin S, Turchetto L, Radrizzani M, Astori G, Schwitter J, Erne P, Zuber M, Auf der Maur C, Jamshidi P, Gaemperli O, Windecker S, Moschovitis A, Wahl A, Buhler I, Wyss C, Kozerke S, Landmesser U, Luscher TF, Corti R (2013) Intracoronary injection of bone marrow-derived mononuclear cells early or late after acute myocardial infarction: effects on global left ventricular function. Circulation 127:1968–1979
Suzuki R, Takizawa T, Negishi Y, Utoguchi N, Maruyama K (2008) Effective gene delivery with novel liposomal bubbles and ultrasonic destruction technology. Int J Pharm 354:49–55
Takahashi M, Li TS, Suzuki R, Kobayashi T, Ito H, Ikeda Y, Matsuzaki M, Hamano K (2006) Cytokines produced by bone marrow cells can contribute to functional improvement of the infarcted heart by protecting cardiomyocytes from ischemic injury. Am J Physiol Heart Circ Physiol 291:H886–H893
Tang Y (2005) Gene therapy for myocardial ischemia using the hypoxia-inducible double plasmid system. Methods Mol Med 112:37–47
Timmers L, Lim SK, Hoefer IE, Arslan F, Lai RC, van Oorschot AA, Goumans MJ, Strijder C, Sze SK, Choo A, Piek JJ, Doevendans PA, Pasterkamp G, de Kleijn DP (2011) Human mesenchymal stem cell-conditioned medium improves cardiac function following myocardial infarction. Stem Cell Res 6:206–214
Timolati F, Ott D, Pentassuglia L, Giraud MN, Perriard JC, Suter TM, Zuppinger C (2006) Neuregulin-1 beta attenuates doxorubicin-induced alterations of excitation-contraction coupling and reduces oxidative stress in adult rat cardiomyocytes. J Mol Cell Cardiol 41:845–854
Toma C, Pittenger MF, Cahill KS, Byrne BJ, Kessler PD (2002) Human mesenchymal stem cells differentiate to a cardiomyocyte phenotype in the adult murine heart. Circulation 105:93–98
Tran C, Damaser MS (2015) Stem cells as drug delivery methods: application of stem cell secretome for regeneration. Adv Drug Deliv Rev 82–83:1–11
Traverse JH, Henry TD, Pepine CJ, Willerson JT, Zhao DX, Ellis SG, Forder JR, Anderson RD, Hatzopoulos AK, Penn MS, Perin EC, Chambers J, Baran KW, Raveendran G, Lambert C, Lerman A, Simon DI, Vaughan DE, Lai D, Gee AP, Taylor DA, Cogle CR, Thomas JD, Olson RE, Bowman S, Francescon J, Geither C, Handberg E, Kappenman C, Westbrook L, Piller LB, Simpson LM, Baraniuk S, Loghin C, Aguilar D, Richman S, Zierold C, Spoon DB, Bettencourt J, Sayre SL, Vojvodic RW, Skarlatos SI, Gordon DJ, Ebert RF, Kwak M, Moye LA, Simari RD (2012) Effect of the use and timing of bone marrow mononuclear cell delivery on left ventricular function after acute myocardial infarction: the TIME randomized trial. JAMA 308:2380–2389
Uccelli A, Moretta L, Pistoia V (2008) Mesenchymal stem cells in health and disease. Nat Rev Immunol 8:726–736
Uemura R, Xu M, Ahmad N, Ashraf M (2006) Bone marrow stem cells prevent left ventricular remodeling of ischemic heart through paracrine signaling. Circ Res 98:1414–1421
Valadi H, Ekstrom K, Bossios A, Sjostrand M, Lee JJ, Lotvall JO (2007) Exosome-mediated transfer of mRNAs and microRNAs is a novel mechanism of genetic exchange between cells. Nat Cell Biol 9:654–659
van Amerongen MJ, Engel FB (2008) Features of cardiomyocyte proliferation and its potential for cardiac regeneration. J Cell Mol Med 12:2233–2244
van Berlo JH, Kanisicak O, Maillet M, Vagnozzi RJ, Karch J, Lin SC, Middleton RC, Marban E, Molkentin JD (2014) c-kit+ cells minimally contribute cardiomyocytes to the heart. Nature 509:337–341
Vlassov AV, Magdaleno S, Setterquist R, Conrad R (2012) Exosomes: current knowledge of their composition, biological functions, and diagnostic and therapeutic potentials. Biochim Biophys Acta 1820:940–948

Wollert KC (2015) Bone marrow mononuclear cell therapy for acute myocardial infarction: we know what we want, but we just don't know how yet. Heart 101:337–338

Wollert KC, Drexler H (2010) Cell therapy for the treatment of coronary heart disease: a critical appraisal. Nat Rev Cardiol 7:204–215

Wollert KC, Meyer GP, Lotz J, Ringes-Lichtenberg S, Lippolt P, Breidenbach C, Fichtner S, Korte T, Hornig B, Messinger D, Arseniev L, Hertenstein B, Ganser A, Drexler H (2004) Intracoronary autologous bone-marrow cell transfer after myocardial infarction: the BOOST randomised controlled clinical trial. Lancet 364:141–148

Wu Y, Chen L, Scott PG, Tredget EE (2007) Mesenchymal stem cells enhance wound healing through differentiation and angiogenesis. Stem Cells 25:2648–2659

Ye J, Yeghiazarians Y (2015) Cardiac stem cell therapy: have we put too much hype in which cell type to use? Heart Fail Rev 20:613–9

Yockman JW, Kastenmeier A, Erickson HM, Brumbach JG, Whitten MG, Albanil A, Li DY, Kim SW, Bull DA (2008) Novel polymer carriers and gene constructs for treatment of myocardial ischemia and infarction. J Control Release 132:260–266

Zebrowski DC, Engel FB (2013) The cardiomyocyte cell cycle in hypertrophy, tissue homeostasis, and regeneration. Rev Physiol Biochem Pharmacol 165:67–96

Zhao ZG, Li WM, Chen ZC, You Y, Zou P (2008) Immunosuppressive properties of mesenchymal stem cells derived from bone marrow of patients with chronic myeloid leukemia. Immunol Invest 37:726–739

Chapter 9
Microvesicles and Exosomes in Local and Distant Communication with the Heart

Sean Davidson

9.1 Introduction

The existence of extracellular vesicles (EVs), such as microvesicles and exosomes has been recognized for some time, but it is only relatively recently that their relevance to cardiovascular health and disease has emerged. In particular, the search for potential mediators of the paracrine benefit of stem cells implanted in the heart has alighted on exosomes as interesting candidate molecules which may deliver protective and regenerative signals to the heart. This chapter will provide a background understanding of exosome and microvesicle biology and practical limitations in their analysis. Data suggesting these vesicles have beneficial effects in the heart will be discussed, as will the evidence for their possible involvement in the beneficial effects of stem cell in the heart.

9.2 Extracellular Vesicles

All eukaryotic cells are bound by a lipid bilayer membrane. If this pinches and buds off from the cell, it will cause the formation of small, bilayer lipid membrane vesicles containing some of the membrane-localized and cytosolic proteins of the "parent" cell. Budding such as this occurs during the regulated process of apoptosis, when the cytoskeleton constricts and the plasma membrane bulges out, causing the familiar "blebbing" appearance of dying cells. Eventually these cells fragment into "apoptotic bodies", which are vesicles with a typical diameter of 1–5 μm.

S. Davidson (✉)
The Hatter Cardiovascular Institute, 67 Chenies Mews, London WC1E 6HX, UK
e-mail: s.davidson@ucl.ac.uk

R. Madonna (ed.), *Stem Cells and Cardiac Regeneration*, Stem Cell Biology and Regenerative Medicine, DOI 10.1007/978-3-319-25427-2_9

However, living cells also release vesicles from the plasma membrane via a process of shedding or budding (Fig. 9.1). These vesicles of a diameter of between 300 nm and 1 μm are called "microvesicles". Even smaller are "exosomes", which can form within endosomes by invagination of the endosome membrane to form internal vesicles of 30–100 nm in diameter. These endosomes containing numerous vesicles are called "multivesicular bodies". When the multivesicular body fuses with the plasma membrane, the exosomes are released from the cell into the extracellular space—a process of exocytosis that is believed to be energy dependent and involve members of the Rab family of small GTPases, although the exact mechanism is not well defined and may be cell type dependent (Raposo and Stoorvogel 2013; Yellon and Davidson 2014; Kowal et al. 2014).

Most, if not all, cell types appear to be capable of releasing exosomes and microvesicles (Raposo and Stoorvogel 2013), but some are more prolific than others. Platelets are some of the most prolific, as discovered by Peter Wolf in 1967, when he realised that "platelet dust" released by activated platelets was actually microvesicles. At the time, however, they were considered to be cellular debris (Wolf 1967). In addition to releasing microvesicles, platelets also produce abundant numbers of exosomes (Heijnen et al. 1999). Exosomes were first discovered by examining the particles released by reticulocytes during their maturation into red blood cells (Johnstone et al. 1987). Again, they were initially thought to be waste products involved in "trash disposal" of proteins such as the transferrin receptor that were no longer required (Johnstone et al. 1987). Dendritic cells were also found to

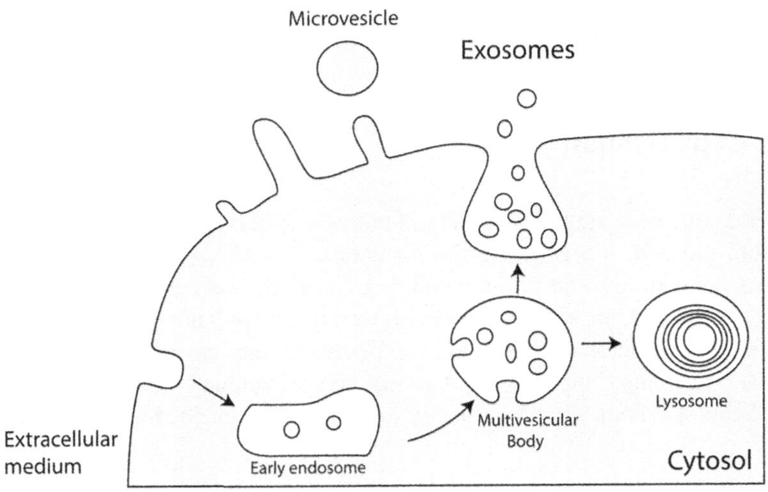

Fig. 9.1 How the main types of EVs are formed. Microvesicles are released from the plasma membrane via a process of shedding or budding. Vesicles form within early endosomes by invagination of the endosome membrane, resulting in what are called multivesicular bodies. Once the multivesicular body fuses with the plasma membrane, the exosomes within are released from the cell into the extracellular space

release exosomes, but these were shown to have a functional effect, since by presenting major histocompatibility complex antigens and generating an immune response, allogeneic donor-derived exosomes could induce immune tolerance to a heart allograft (Peche et al. 2003). In 2001, tumour cells were shown to release large numbers of exosomes which, when taken up by dendritic cells, could initiate a cytotoxic T-cell response (Wolfers et al. 2001). Exosomes have remained an intensive area of research in the cancer field, although the general opinion is now that exosomes aid the tumour cells in escaping immune surveillance by suppressing the immune response (Zhang and Grizzle 2011), and may facilitate metastasis (Hood et al. 2011).

In the cardiovascular system, exosomes have been isolated from primary adult rat cardiomyocytes (Gupta and Knowlton 2007; Wang et al. 2014) as well as the HL-1 mouse atrial cardiomyocyte cell line (Genneback et al. 2013; Waldenstrom et al. 2012). Matrix vesicles, which are secreted by vascular smooth muscle cells and are involved in initiating vascular calcification, were recently shown to be exosomes (Kapustin et al. 2015). Endothelial cells appear to release both exosomes and microvesicles (Deregibus et al. 2007). The production of exosomes is difficult to confirm in vivo. Vesicle-like structures resembling exosomes have been identified in electron micrographs of mouse heart (Barile et al. 2012), but in order to confirm their identity as exosomes and not unrelated vesicles or cross-sections of finger-like projections, for example, it would be necessary to immunolabel them with antibodies against known exosomal markers proteins. It is also very difficult to determine cell of origin in static images. Vascular smooth muscle cells secrete matrix vesicles that initiate vascular calcification. These vesicles were recently identified as exosomes containing some proteins to osteoblast-derived matrix vesicles (Kapustin et al. 2015), demonstrating an intriguing role for exosomes in vascular pathology. Finally, all body fluids contain large numbers of exosomes and microvesicles (Yellon and Davidson 2014; Thery et al. 2006). In the case of blood, the majority of EVs originate from platelets (Smalley et al. 2007), though erythrocytes, endothelium, lymphocytes or other blood cells also contribute.

9.3 Isolation of Extracellular Vesicles

In all studies of EVs the isolation method used is a primary consideration. This is particularly the case for exosomes, for which there is as yet no "perfect" method of purification. In fact, the International Society of Extracellular Vesicles (ISEV) released a position statement in 2014 recognizing that most isolation methods result in a mixture of different types of EVs (Lotvall et al. 2014). This is partly a consequence of the functional definition of exosomes as "vesicles that are released from multivesicular bodies"—once released from the cell there is no unique method to identify their origin. It is also experimentally challenging to use size to separate exosomes from microvesicles, which are only slightly larger, and from other vesicles, protein aggregates, and lipoproteins, which can overlap in size (Lotvall et al. 2014; Thery et al. 2006). Density gradient centrifugation is often used to separate different vesicle types based on their buoyant density (Lotvall et al. 2014; Thery et al. 2006). This can be useful for separating them from protein aggregates which do not float, but does not completely eliminate lipoproteins (HDL and LDL) (Lotvall et al. 2014). Protein markers are commonly used to identify exosomes and microvesicles in preparations, and these are discussed under the following section. In general, cardiovascular researchers will be interested in isolating exosomes either from in vitro cultured cells, or from blood. The critical issues for each are somewhat different and are addressed separately.

With regard to the isolation of vesicles from in vitro cultured cells, most cell lines are grown in the presence of serum, which contains enormous quantities of exosomes. It is therefore essential to obtain specially filtered, certified exosome-free serum, or to prepare one's own by overnight ultracentrifugation of the serum at 100,000 g or higher. Alternatively, it may be possible to grow some cell lines in the absence of serum during the experiment, although care must be taken that the cells remain healthy and do not undergo apoptosis, which will cause the release of apoptotic bodies that can co-purify with microvesicles or exosomes. The most efficient way of collecting exosomes from large volumes of cell medium is serial centrifugation. Sometimes, a filtration step is used first to concentrate the exosomes, however this is prone to artifactual generation of exosome-sized vesicles by fragmentation of the larger vesicles (Lotvall et al. 2014). In the process of serial centrifugation, the supernatant is collected and damaged cells and larger vesicles are removed by centrifugation at 10,000 for 30 min. The supernatant is then ultracentrifuged, typically at 100,000 g for 60 min to pellet exosomes (Hergenreider et al. 2012; Thery et al. 2006). As mentioned, the isolated exosome-rich fraction may be further purified by density gradient centrifugation on sucrose or iodixanol to enrich exosomes which have a density of between 1.13 and 1.19 g/mL (Yuana et al. 2014; Thery et al. 2006). This fraction is then diluted in buffer and re-centrifuged to remove the sucrose or iodixanol. One alternative method is to separate exosomes by size-exclusion chromatography (Boing et al. 2014). Although effective, this is less widely used due to lower throughput. A further technique is to use antibody affinity capture on beads or columns, although this requires that one is certain of the antigen marking the vesicles of interest, and the yield is much lower. Several companies now market reagents that can be used to rapidly precipitate exosomes from culture medium (or serum), the original being "Exoquick™". Although yield is high, the purity is suspect (Van Deun et al. 2014), since it is difficult to wash the isolated exosome pellet washed without resorting to ultracentrifugation.

The isolation of EVs from blood is further complicated by pre-analytical considerations (Yuana et al. 2011). These are mostly designed to prevent or minimize the activation of platelets, which will otherwise release enormous numbers of microvesicles and exosomes which may influence the results. The clotting step involved in serum preparation causes massive platelet activation, which can be useful if large numbers of platelet vesicles are desired. However it is necessary to use plasma to obtain a selection of EVs representative of blood EVs (Yuana et al. 2011). Best practise is to use a relatedly large bore butterfly needle and draw blood into citrated Vacutainer® tubes, discarding the first tube. After mixing gently by inversion, the tube is centrifuged for at 1600×*g* for 20 min at room temperature (to minimize platelet activation) and the platelet-free plasma is carefully collected from the top. The standard protocol (Thery et al. 2006) is then to centrifuge the plasma at 10,000×*g* for 30 min at room temperature to pellet microvesicles. The supernatant containing exosomes is removed and ultracentrifuged at 100,000×*g* for 90 min at 4 °C. The supernatant is removed and the pellet resuspended in Ca^{2+} and Mg^{2+} free PBS containing citrate, and ultracentrifuged again. After a final wash, the pellet (which is not usually visible) can be vigorously resuspended for analysis of exosomes. Similar to exosomes isolated from tissue culture, density gradient purification may be performed to improve purity from protein aggregates, although contamination with high-density lipoproteins is very difficult to eliminate (Yuana et al. 2014).

Although some publications have used precipitation reagents such as Exoquick™ to isolate exosomes from blood, the standard protocol was designed for use with serum and must be adapted by the addition of Thromboplastin D in order to remove coagulative proteins and enable the use of plasma. Furthermore, the difficulty of washing the pellet containing exosomes is a major issue, and since blood contains very high concentrations of proteins, microvesicles, etc. some of these will remain in the final isolate. Despite these issues, some researchers have successfully used these methods to identify signature differences in exosome content between groups of individuals, which may be useful for diagnostic purposes. However, it must be taken into consideration that the exosomal origin of these samples remains unproven.

A promising new technique combining the convenience of precipitation and a premade kit with the improved separation of columns is the Exo-Spin™ kit, which appear to provide a convenient and reproducible means of eliminating >95 % of non-vesicular protein from biological fluid samples such as plasma (Welton et al. 2015).

9.4 Characterization of Exosomes and Microvesicles

Several approaches are necessary to unambiguously identify exosomes. Since their diameter is smaller than the wavelength of light, direct imaging of individual exosome using standard microscopy is impossible. Furthermore, they lie well below the detection threshold of flow cytometry (Nolte-'t Hoen et al. 2012). Flow cytometry is feasible for analysis of microvesicles but specialized modifications to flow cytometer equipment are necessary to optimize detection of particles at the lower range of size (Nolte-'t Hoen et al. 2012).

The most direct means of visualizing EVs is by transmission electron micrography after drying and uranyl acetate staining (Thery et al. 2006). Exosomes and other vesicles exhibit a shape that has been called "cup-shaped" as the vesicles collapse when dried (Fig. 9.2a) (Thery et al. 2006; Raposo and Stoorvogel 2013). Contaminating objects can include lipoproteins (which appear spherical), protein

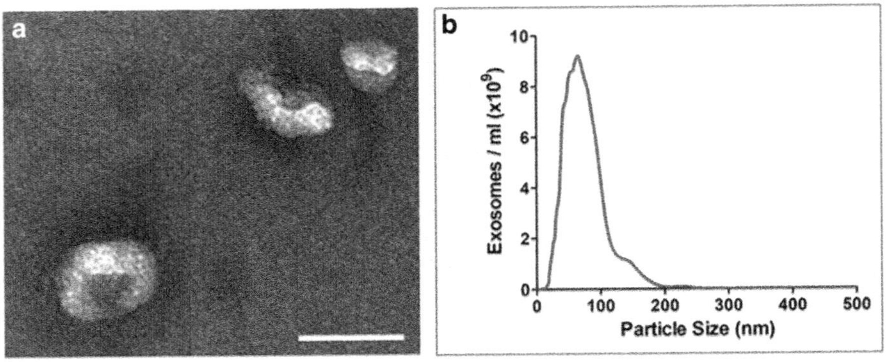

Fig. 9.2 (**a**) Exosomes isolated from human plasma by serial centrifugation and visualized by electron micrography. Scale = 100 nm. (**b**) Size distribution of human plasma exosomes determined by nanoparticle tracking analysis. The majority are 50–100 nm although some larger particles are also present

aggregates, cellular organelles or viruses. Sometimes the background of the electron micrograph grid itself can resemble objects of the appropriate size range for exosomes, so it is useful to include a control grid without exosomes for comparison. Positive identification by electron micrograph is an important step in confirming the presence of exosomes and calculating the size distribution of a vesicle population.

Vesicle size can also be determined using dynamic light scattering (DLS) or nanoparticle tracking analysis (NTA) (Fig. 9.2b) (Dragovic et al. 2011). DLS can be effective but when a vesicle preparation contains populations of different sizes NTA is more appropriate as it is able to measure the size of individual particles.

Exosomes contain particular protein markers as a consequence of their release mechanism from cells, and these are often used to aid their identification (or utilized for vesicle purification by antibody affinity). Foremost amongst these proteins are the tetraspanins, particularly CD9, CD63, CD81, although it must be recognized that other vesicles can express these molecules as well (Kowal et al. 2014). It may be possible to detect these proteins by Western blot analysis, although non-reducing gels should be used because the tetraspanins are heavily glycosylated (Thery et al. 2006). If it is difficult to obtain sufficient protein for a Western blot, an effective alternative is to adhere the exosomes onto 4 μm diameter latex beads, incubate them with fluorescent antibodies against tetraspanins or other marker proteins, and measure the fluorescence of individual beads separated by flow cytometry (Thery et al. 2006). Modern flow cytometers are quite capable of analysing most microvesicles (Ayers et al. 2014), although the smallest microvesicles might be beyond the range of detection without careful optimization of the flow cytometer.

It is recommended to use several of the above approaches in order to confirm the identity and origin of the EVs.

9.5 Extracellular Vesicles Can Transfer Proteins and RNA

EVs have been demonstrated to possess a number of interesting capabilities including the ability to deliver proteins and RNA to cells. There is some evidence that this also occurs in vivo, despite this being much more difficult to prove. One of the first proteins demonstrated to be transferred to recipient cells in vitro was tissue factor (TF). TF was released from platelets within microvesicles that were subsequently incorporated into the membranes of monocytes and other cells where TF exerts its biological effects (Del Conde et al. 2005; Scholz et al. 2002). An important paper demonstrated that microvesicles from tumour cells could transfer a truncated, oncogenic form of the epidermal growth factor receptor between cells, leading to activation of transforming signalling pathways (MAPK and Akt) and horizontal propagation of oncogenes and their associated transforming phenotype (Al-Nedawi et al. 2008).

Exosomes are usually found to contain primarily cytosolic and membrane-localized proteins (Mathivanan et al. 2010). Certain proteins are found almost universally in exosomes due to their being part of the packaging and/or release pathway. These include the proteins Alix, TSP101, HSP70 and tetraspanin family proteins CD9, CD63, CD81, etc., which are therefore often used as markers proteins to indicate the presence of exosomes, as described in the previous section. The degree of sorting of proteins into exosomes and microvesicles appears to be greater when the proteins contain domains that link them to the plasma membrane (Shen et al. 2011), presumably as this makes them more likely to be incorporated into EVs as the plasma membrane buds off or invaginates to produce multivesicular bodies. There is some evidence that the protein and RNA content of exosomes depends on the cellular state (de Jong et al. 2012).

Similar to microvesicles, exosomes can not only deliver signalling proteins to cells, but be taken up by other cells, incorporating the protein into the recipient membrane. For example, the Notch ligand Delta-like 4 (Dll4) is incorporated into endothelial exosomes which can transfer the Dll4 protein to other endothelial cells and incorporate it into their cell membrane, resulting in an inhibition of Notch signalling and altered angiogenesis (Sheldon et al. 2010). Of particular interest are the examples in which exosomes have been shown to deliver protective proteins to recipient cells. αB crystallin was shown to be secreted from human retinal pigment epithelium in exosomes, and taken up by adjacent photoreceptors in mouse retinal explants exposed to oxidative stress (Sreekumar et al. 2010). An important theoretical consideration is the extent to which transfer of a relatively small number of copies of a protein can be realistically expected to have a physiological effect in the recipient cell in vivo. In this respect, it is easier to imagine transfer of small quantities of mRNA or miRNA being capable of having more dramatic effects.

In a seminal paper, Valadi et al, were first to show that exosomes can transfer mRNA and miRNA between cells (Valadi et al. 2007). This study used mast cells to demonstrate the transfer of functional mRNAs that were subsequently translated in the recipient cells. In an important control experiment, exosomes were pre-treated with RNAse and trypsin, to demonstrate that mRNA was protected within the vesicles and not simply associated or co-purified. Tumour-derived microvesicles were also shown to transfer mRNA to recipient cells that could be translated (Skog et al. 2008).

The profile of miRNAs contained within exosomes appears to differ according to the cell type of origin. For example, C2C12 myoblasts and C2C12 differentiated into myotubes released exosomes with different miRNA profiles (Forterre et al. 2014). Interestingly, the miRNA profile was different from that of the donor cells, indicating that there is selective packaging of miRNA into exosomes (Forterre et al. 2014). The mechanism of selective sorting of miRNA into exosomes is only beginning to be unravelled, but is believed to involve recognition of particular sequence motifs by sumoylated heterogeneous nuclear ribonucleoprotein A2B1 (hnRNPA2B1) (Villarroya-Beltri et al. 2013). The exosomes secreted by C2C12 myotubes were functional, being able to suppress expression of Sirt1 when taken up by myoblasts, potentially modulating metabolic homeostasis and contributing to the commitment of myoblasts during differentiation (Forterre et al. 2014).

Cardiac fibroblasts secrete miRNA-enriched exosomes, including miR-21-3p—a passenger strand miRNA which normally undergoes intracellular degradation (Bang et al. 2014). However, when these exosomes were taken up by neonatal cardiomyocytes in co-culture, the cardiomyocytes increased in size indicating a hypertrophic response (Bang et al. 2014). Thus it appears that different cells in the heart may communicate to each other via exosomes. A further significant exchange of miRNAs between cells of the cardiovascular system was shown to occur between endothelial cells and smooth muscle cells. In this case, EVs were secreted by endothelial cells that had been stimulated by shear stress, which is known to be atheroprotective (Hergenreider et al. 2012). The EVs were taken up into smooth muscle cells in co-culture, delivering miR-143/145 and controlling the expression of target genes (Hergenreider et al. 2012). They also reduced atherosclerotic lesion formation in the aorta of ApoE(−/−) mice (Hergenreider et al. 2012). Since a maximum centrifugation speed of 20,500×*g* was used to pellet EVs, and the size range of most of the vesicles on electron micrographs ranged between 60 and 130 nm, they were referred to conservatively as "extracellular vesicles", but presumably contained a mix of exosomes and microvesicles.

In view of the RNA content of EVs which is related to the cell type of origin, and can alter in pathological settings, they have become an attractive source of biomarkers for profiling and identification of disease markers (Cheng et al. 2014; Jansen et al. 2014), as has been reviewed elsewhere (Gaceb et al. 2014).

9.5.1 Extracellular Vesicles Can Transfer Proteins and RNA

EVs have been demonstrated to possess a number of interesting capabilities including the ability to deliver proteins and RNA to cells. There is some evidence that this also occurs in vivo, despite this being much more difficult to prove. One of the first proteins demonstrated to be transferred to recipient cells in vitro was tissue factor (TF). TF was released from platelets within microvesicles that were subsequently incorporated into the membranes of monocytes and other cells where TF exerts its biological effects (Del Conde et al. 2005; Scholz et al. 2002). An important paper demonstrated that microvesicles from tumour cells could transfer a truncated, oncogenic form of the epidermal growth factor receptor between cells, leading to activation of transforming signalling pathways (MAPK and Akt) and horizontal propagation of oncogenes and their associated transforming phenotype (Al-Nedawi et al. 2008).

Exosomes are usually found to contain primarily cytosolic and membrane-localized proteins (Mathivanan et al. 2010). Certain proteins are found almost universally in exosomes due to their being part of the packaging and/or release pathway. These include the proteins Alix, TSP101, HSP70 and tetraspanin family proteins CD9, CD63, CD81, etc., which are therefore often used as markers proteins to indicate the presence of exosomes, as described in the previous section. The degree of sorting of proteins into exosomes and microvesicles appears to be greater when the proteins contain domains that link them to the plasma membrane (Shen et al. 2011), presumably as this makes them more likely to be incorporated into EVs as the plasma membrane buds off or invaginates to produce multivesicular bodies. There is some evidence that the protein and RNA content of exosomes depends on the cellular state (de Jong et al. 2012).

Similar to microvesicles, exosomes can not only deliver signalling proteins to cells, but be taken up by other cells, incorporating the protein into the recipient membrane. For example, the Notch ligand Delta-like 4 (Dll4) is incorporated into endothelial exosomes which can transfer the Dll4 protein to other endothelial cells and incorporate it into their cell membrane, resulting in an inhibition of Notch signalling and altered angiogenesis (Sheldon et al. 2010). Of particular interest are the examples in which exosomes have been shown to deliver protective proteins to recipient cells. αB crystallin was shown to be secreted from human retinal pigment epithelium in exosomes, and taken up by adjacent photoreceptors in mouse retinal explants exposed to oxidative stress (Sreekumar et al. 2010). An important theoretical consideration is the extent to which transfer of a relatively small number of copies of a protein can be realistically expected to have a physiological effect in the recipient cell in vivo. In this respect, it is easier to imagine transfer of small quantities of mRNA or miRNA being capable of having more dramatic effects.

In a seminal paper, Valadi et al, were first to show that exosomes can transfer mRNA and miRNA between cells (Valadi et al. 2007). This study used mast cells to demonstrate the transfer of functional mRNAs that were subsequently translated in the recipient cells. In an important control experiment, exosomes were pre-treated with RNAse and trypsin, to demonstrate that mRNA was protected within the vesicles and not simply associated or co-purified. Tumour-derived microvesicles were also shown to transfer mRNA to recipient cells that could be translated (Skog et al. 2008).

The profile of miRNAs contained within exosomes appears to differ according to the cell type of origin. For example, C2C12 myoblasts and C2C12 differentiated into myotubes released exosomes with different miRNA profiles

(Forterre et al. 2014). Interestingly, the miRNA profile was different from that of the donor cells, indicating that there is selective packaging of miRNA into exosomes (Forterre et al. 2014). The mechanism of selective sorting of miRNA into exosomes is only beginning to be unravelled, but is believed to involve recognition of particular sequence motifs by sumoylated heterogeneous nuclear ribonucleoprotein A2B1 (hnRNPA2B1) (Villarroya-Beltri et al. 2013). The exosomes secreted by C2C12 myotubes were functional, being able to suppress expression of Sirt1 when taken up by myoblasts, potentially modulating metabolic homeostasis and contributing to the commitment of myoblasts during differentiation (Forterre et al. 2014).

Cardiac fibroblasts secrete miRNA-enriched exosomes, including miR-21-3p—a passenger strand miRNA which normally undergoes intracellular degradation (Bang et al. 2014). However, when these exosomes were taken up by neonatal cardiomyocytes in co-culture, the cardiomyocytes increased in size indicating a hypertrophic response (Bang et al. 2014). Thus it appears that different cells in the heart may communicate to each other via exosomes. A further significant exchange of miRNAs between cells of the cardiovascular system was shown to occur between endothelial cells and smooth muscle cells. In this case, EVs were secreted by endothelial cells that had been stimulated by shear stress, which is known to be atheroprotective (Hergenreider et al. 2012). The EVs were taken up into smooth muscle cells in co-culture, delivering miR-143/145 and controlling the expression of target genes (Hergenreider et al. 2012). They also reduced atherosclerotic lesion formation in the aorta of ApoE(−/−) mice (Hergenreider et al. 2012). Since a maximum centrifugation speed of 20,500 g was used to pellet EVs, and the size range of most of the vesicles on electron micrographs ranged between 60 and 130 nm, they were referred to conservatively as "extracellular vesicles", but presumably contained a mix of exosomes and microvesicles.

In view of the RNA content of EVs which is related to the cell type of origin, and can alter in pathological settings, they have become an attractive source of biomarkers for profiling and identification of disease markers (Cheng et al. 2014; Jansen et al. 2014), as has been reviewed elsewhere (Gaceb et al. 2014).

9.6 Exosomes as a Potential Therapy for Cardiovascular Disease

Coronary artery disease is a major challenge facing health care systems around the world. In 2004, ischaemic heart disease was responsible for 7.2 million deaths, making it the leading cause of death. Myocardial infarction is a major cause of mortality and morbidity in patients with coronary artery disease. Heart failure often develops subsequently, depending on the extent of myocardial infarction. Therefore, treatment modalities that protect the heart from injury can be expected to result in major improvements not only in early mortality but on-going morbidity. Currently early reperfusion (restoration of normal coronary flow) remains the mainstay of treatment for those suffering an acute myocardial infarction. However, reperfusion also causes injury to heart muscle in what is termed "lethal reperfusion injury" (Hausenloy and Yellon 2013). Final infarct size in patients with ST elevation myocardial infarction predicts long-term clinical outcome (Lonborg et al. 2013). Therefore, finding ways to minimize reperfusion injury is one approach that would be expected to improve the overall, long-term management of patients presenting with myocardial infarction (Braunwald and Kloner 1985; Piper et al. 1998; Hausenloy and Yellon 2013). Another theoretical approach would be to replace the contractile cardiomyocytes that were lost, by the use of stem cell therapy. As discussed elsewhere in this book, however, the results of this approach have been largely disappointing. On the other hand, some improvements in cardiac function have been observed after stem cell therapy, even in the absence of new cardiomyocytes formation. Since some benefit could also be observed experimentally after injection of conditioned medium from stem cells, the "paracrine" hypothesis developed in which it is proposed that engrafted stem cells release cytokines, growth factors and other proteins which act on cardiomyocytes and other myocardial proteins to confer improvements in function. Stem cells release a host of different factors ("the secretome"), many of which are pro-survival and proangiogenic (Makridakis et al. 2013). While many of these are released via the usual protein secretion pathways, the possibility was raised that they might be released together associated with exosomes, enabling a co-ordinated, and perhaps directed response to multiple factors.

The first study to examine whether exosomes released from stem cells are cardioprotective was performed in 2010 using an acute model of myocardial infarction. Timmers et al. had previously showed that conditioned medium from human ESC-derived mesenchymal stem cells (ESC-MSC) reduced infarct size in isolated, perfused mouse hearts subject to ischaemia and reperfusion (Timmers et al. 2007). The same group then purified exosomes from the conditioned medium by HPLC size-exclusion fractionation, and confirmed that these exosomes were sufficient to confer cardioprotection both in vitro and in vivo (Lai et al. 2010a). Long term protection with improved function after 28 days was observed after injection of at least 4 μg/kg into mice via the tail vein, just before reperfusion (Arslan et al. 2013). Interestingly, it appears that the exosomes had to be intact to induce protection, since they were no longer protective after homogenization (Arslan et al. 2013), although whether exosomes are disrupted after homogenization is not clear and was not demonstrated. Exosome administration caused an increase in the activity of cardioprotective kinases Akt and GSK3α/β after 1 h, and these remained activated until the day following treatment (Arslan et al. 2013). These kinases are known to be highly cardioprotective (Hausenloy et al. 2005). Significantly, there was also a reduction in myocardial oxidative stress, and local and systemic inflammation (Arslan et al. 2013). Exosomes isolated from MSC cells overexpressing GATA4 have also been shown to restore cardiac contractile function and reduce infarct size when injected intramyocardially at the time of infarction in a rat model (Yu et al. 2014). An increase in miR-19a was detected in the treated hearts, although it is difficult to be certain that this originated from the MSC exosomes and was not a response to treatment with exosomes (Yu et al. 2014). In turn, levels of PTEN, a predicted target of miR-19a, were reduced and Akt and ERK activation was increased (Yu et al. 2014). As an alternative cell source for large-scale production of exosomes which does not require ES cells, MSC cultures were developed from limb, kidney and liver tissues of three first trimester aborted foetuses. Despite absence of pluripotency-associated markers, these MSC were also an excellent source of cardioprotective exosomes (Lai et al. 2010b). Of interest, the protective pathways activated by exosomes appear to be applicable to other organs, since MSC exosomes (from human umbilical cord) have also been shown to protect against cisplatin-induced renal oxidative stress and apoptosis via ERK (Zhou et al. 2013), and to promote functional recovery after stroke in rats (Xin et al. 2013). Furthermore, it may be that protection is not limited to exosomes, as microvesicles derived from human adult mesenchymal stem cells also protected against ischaemia and reperfusion kidney injury (Gatti et al. 2011).

MSC are not the only type of stem cell that have been shown to release exosomes with beneficial cardiovascular effects. Intramyocardial injection of cardiac progenitor cells (CPCs) isolated from adult hearts improved cardiac function in animal models of myocardial infarction (Messina et al. 2004; Smith et al. 2007), although paracrine effects were found to contribute to the majority of the effect (Chimenti et al. 2010). Subsequently, exosomes from murine CPCs were found to reduce apoptosis when injected intramyocardially during ischaemia (Chen et al. 2013). In this study, however, exosomes were isolated by PEG precipitation (Chen et al. 2013), and therefore the more appropriate control would have been injection with PEG alone. Another study used several alternative methods of isolation of CPC-EVs: Exoquick™ precipitation, ultracentrifugation, or size-exclusion, column purification (Barile et al. 2014). Although more of tetraspanin marker proteins CD63 and CD81 were measured using Exoquick™, the three techniques were not rigorously compared, and as mentioned above, the Exoquick™ technique has some drawbacks—particularly the inability to wash the pellet to remove the Exoquick™ solution from the exosomes. Injection of these CPCs-EVs into the hearts of rats subject to permanent coronary artery ligation reduced cardiomyocyte apoptosis and scar size, increased the amount of viable tissue in the infarct area, increased blood vessel density, and prevented the loss of impairment of ventricular function between day 2 and day 7 ((Barile et al. 2014). In contrast, exosomes isolated from normal human dermal fibroblasts exhibited no benefit (Barile et al. 2014). Gray et al. used ultracentrifugation to purify exosomes from CPCs, and found that the rate of secretion of exosomes was increased after 3–12 h hypoxia (Gray et al. 2015). Intramyocardial injection of 5 μg/kg of the exosomes produced after hypoxia improved cardiac function and reduced fibrosis 21 days after ischaemia and reperfusion injury in rats (Gray et al. 2015). Interestingly, however, there was no benefit with exosomes from normoxic CPCs (Barile et al. 2014). The exosomes released after hypoxia had an altered miRNA content, and co-regulated miRNA with a beneficial profile were identified (Gray et al. 2015). Although cardiac endothelial cells and fibroblasts took up fluorescently stained exosomes in vitro, uptake was minimal in primary rat cardiomyocytes (Gray et al. 2015), suggesting either that they deliver miRNA directly to the former cells types, or that they interact with surface receptors on cardiomyocytes without delivering miRNA intracellularly. Thus, the exact mechanism of functional benefit conferred by exosomes remains unclear.

Various avenues are being explored to increase the survival and engraftment of stem cells once they have been injected into the heart. Recently, CPCs were co-delivered into ischemic mouse myocardium with a nonviral minicircle plasmid carrying HIF1, a transcription factor that mediates adaptive responses to ischemia (Ong et al. 2014). As intended, this increased the survival of CPCs after injection, and also significantly improved cardiac function 6 weeks later (Ong et al. 2014). The authors hypothesized that HIF1 was being expressed by cardiac endothelial cells, thereby modulating the host microenvironment such that CPC survival was increased. Exosomes were purified from endothelial cells overexpressing HIF1 and found to have a higher content of miR-126 and miR-210 (Ong et al. 2014). In in vitro experiments, these exosomes were taken up by CPCs, leading to activated pro-survival kinases and to a switch towards glycolysis in recipient CPCs, giving them increased tolerance when subjected to in vitro hypoxic stress (Ong et al. 2014). Antagomirs to these miRs blocked the protective effects of exosomes (Ong et al. 2014). This suggests the interesting possibility that endothelial cells can support CPS survival by exosomal transfer of miRNA.

Bone marrow CD34+ stem cells are another type of stem cell that been associated with reduced angina, improved exercise time, and reduced amputation rates when delivered to ischaemic myocardium. Exosomes isolated from the conditioned medium of CD34+ cells exerted a proangiogenic paracrine activity, both in vitro and in vivo when tested using Matrigel plug and corneal assays (Sahoo et al. 2011). In contrast, CD34-mononuclear cells did not have this effect. Mackie et al. attempted to enhance the angiogenic quality of CD34 cells by genetically modifying them to express the sonic hedgehog (Shh) protein, a developmental morphogenic protein that is also proangiogenic (see Sect. 10.6.1). CD34+ were first engineered to express Shh. When injected into the infarct border zone in mice after myocardial infarction, infarct size was reduced and border zone capillary density was increased 4 weeks later (Mackie et al. 2012). This was paralleled by improved ventricular dilation and cardiac function. In vitro studies in cells were performed to demonstrate that Shh was released from the CD34+Shh cells in exosomes, and could be transferred to recipient cells and (modestly) activate transcription. When exosomes were isolated from conditioned medium of CD34 + Shh cells using differential ultracentrifugation they were found to contain Shh (Mackie et al. 2012). When injected into mice, were also found to be protective (Mackie et al. 2012). In contrast, CD34+ cells or their exosomes showed no benefit (Mackie et al. 2012).

In all of the above studies, it is important to recall that stem cells also secrete numerous other proteins that are not associated with vesicles. These include pro-inflammatory, anti-inflammatory and pleiotropic cytokines, chemokines, angiogenic factors, growth factors, growth factor-binding proteins, extracellular matrix proteins and extracellular matrix remodelling enzymes (Makridakis et al. 2013; Skalnikova et al. 2011). Many of these, particularly the growth factors and chemokines have been shown to be cardioprotective (Hausenloy and Yellon 2009; Hausenloy et al. 2010), or, along with angiogenic factors, to improve long term recovery after infarc-

Table 9.1 Potential positives and negatives of stem cells and exosomes as therapeutic agents for cardiovascular disease

Exosomes	Stem cells
✔ Mass production from cultured cells theoretically possible	✖ Need to inject large numbers since only a small percentage engraft. Cells may drift genetically while being expanded in culture
✖ Unknown long-term effects	✖ Potentially tumourigenic
✔ Immune suppression should not be necessary	✖ Immune suppression required if foreign cells are used
✔ Transient response only	✔ Long-term engraftment possible
✖ Long-term effects may require repeat administration	✖ But once engrafted, difficult to remove if necessary
✔ Can be easily frozen and stored	✖ Require specialized freezing facilities

tion. Thus, there is still great interest in using stem cells as localized "factories" for production of trophic factors to maintain or improve cardiac function. On the other hand, there are some clear potential benefits to the use of exosomes as delivery vehicles (Table 9.1). Foremost amongst these, since living cells are not injected, is the absence of risk of transformation and tumourigenicity associated with the injection of stem cells. Since injected exosomes are only present transiently, and are of low immunogenicity, there should not be any need for immunosuppression. However, this does mean that there is only a transient response, and repeated administration of exosomes may be required for prolonged benefit. As there have not yet been any clinical trials administering exosomes to humans, however, most of these considerations are hypothetical at present.

The extent to which EVs from different cell types have different functions and activities is still under investigation. In some of the studies above using EVs isolated from stem cells, EVs from other cell types such as fibroblasts were used as negative controls and found not to have beneficial cardiovascular effects in the models used. Exosomes from CD34+ cells were only protective if isolated from cells overexpressing Shh (Mackie et al. 2012). This data must be reconciled with the fact that there are huge numbers of EVs present in the blood of all individuals, which could potentially deliver miRNA or receptor-ligand mediated signals to the heart. This possibility was addressed by isolating plasma exosomes from healthy individuals using differential ultracentrifugation and testing whether they were cardioprotective in rat models of ischaemia and reperfusion in vitro and in vivo (Vicencio et al. 2015). Indeed, exosomes from plasma were strongly cardioprotective, reducing infarct size after injection of all of the exosomes isolated from a donor rat into a recipient rat (Vicencio et al. 2015). Plasma exosomes were similarly protective in an isolated perfused rat heart model (Vicencio et al. 2015). Remote ischaemic preconditioning (RIC) is a technique in which several short episodes of non-lethal ischaemia and reperfusion are applied to the limb by occluding the blood flow using a tourniquet or other means (Hausenloy and Yellon 2008). In animal models this has been shown to induce protection against lethal myocardial ischaemia and reperfusion injury, via a mechanism that involves the transmission of a humoral blood factor to the heart to activate protective kinase signalling pathways (Hausenloy and Yellon 2008). As yet, the identity of the humoral factor is unclear although evidence for several candidates has been presented, including SDF-1α and Il-10 (Davidson et al. 2013; Cai et al. 2012). As vehicles able to deliver multiple signals between cells, EVs were proposed as possible candidates for carriers of the cardioprotective factor released by RIC (Yellon and Davidson 2014). A study by Giricz et al. suggested that this may be the case, since RIC was not effective when EVs were removed from medium containing the factor (Giricz et al. 2014). More recently, the numbers of EVs in the blood was shown to be significantly increased after RIC in both rats and healthy human volunteers (Vicencio et al. 2015). Experiments performed using primary adult rat cardiomyocytes demonstrated a dose dependency. The EVs released after RIC were found to be marginally, but not significantly more protective in this model (Vicencio et al. 2015), suggesting that the increase in EV number was not sufficient to explain the mechanism of RIC. However, the fact that plasma EVs in themselves were cardioprotective is important and may suggest that they signal continuously to the heart, modulating protective state. This is difficult to prove without specific methods of inhibiting EV release in vivo. This study also examined the mechanism of protection using inhibitors and Western blot analysis and identified HSP70 in the exosome membrane as essential for the protection (Vicencio et al. 2015). Protection was blocked by inhibitors of TLR4, or ERK1/2, leading to the proposal of a model in which HSP70 is recognized by TLR4, which activates ERK1/2, p38MAPK and downstream phosphorylation of small heat shock protein HSP27 (Vicencio et al. 2015). TLR4 is part of the innate immune system, and strong activation by its ligands from bacteria leads to a cell damage response and can cause cell death. However, mild activation appears to be protective (Mathur et al. 2011; Zhang et al. 2013).

9.7 Conclusion and Perspectives

In conclusion, exosomes and microvesicles represent an exciting new frontier in cardiovascular research. Although data suggests that they possess an important role in cell-to-cell communication, this remains to be established and defined using rigorously purified vesicle populations for which techniques are yet to be established. Furthermore, crucial experiments to define their role in vivo require the ability to inhibit EV production and/or uptake in vivo—techniques which are not yet available. However, the intense interest in this area of research, and the eagerness to identify the mechanism of potential paracrine benefit by stem cells are two factors likely to contribute to the achievement of remarkable and rapid advancements in this field.

Acknowledgements This work was funded by a grant from the Medical Research Council [MR/K002066/1] and the British Heart Foundation [RG/08/015/26411]. I am grateful for the ongoing support of Prof Derek Yellon, and co-workers at the Hatter Cardiovascular Institute.

References

Al-Nedawi K, Meehan B, Micallef J, Lhotak V, May L, Guha A, Rak J (2008) Intercellular transfer of the oncogenic receptor EGFRvIII by microvesicles derived from tumour cells. Nat Cell Biol 10(5):619–624. doi:10.1038/ncb1725

Arslan F, Lai RC, Smeets MB, Akeroyd L, Choo A, Aguor EN, Timmers L, van Rijen HV, Doevendans PA, Pasterkamp G, Lim SK, de Kleijn DP (2013) Mesenchymal stem cell-derived exosomes increase ATP levels, decrease oxidative stress and activate PI3K/Akt pathway to enhance myocardial viability and prevent adverse remodeling after myocardial ischemia/reperfusion injury. Stem Cell Res 10(3):301–312. doi:10.1016/j.scr.2013.01.002

Ayers L, Harrison P, Kohler M, Ferry B (2014) Procoagulant and platelet-derived microvesicle absolute counts determined by flow cytometry correlates with a measurement of their functional capacity. J Extracell Vesicles 3. doi:10.3402/jev.v3.25348

Bang C, Batkai S, Dangwal S, Gupta SK, Foinquinos A, Holzmann A, Just A, Remke J, Zimmer K, Zeug A, Ponimaskin E, Schmiedl A, Yin X, Mayr M, Halder R, Fischer A, Engelhardt S, Wei Y, Schober A, Fiedler J, Thum T (2014) Cardiac fibroblast-derived microRNA passenger strand-enriched exosomes mediate cardiomyocyte hypertrophy. J Clin Invest 124(5):2136–2146. doi:10.1172/JCI70577

Barile L, Gherghiceanu M, Popescu LM, Moccetti T, Vassalli G (2012) Ultrastructural evidence of exosome secretion by progenitor cells in adult mouse myocardium and adult human cardiospheres. J Biomed Biotechnol 2012:354605. doi:10.1155/2012/354605

Barile L, Lionetti V, Cervio E, Matteucci M, Gherghiceanu M, Popescu LM, Torre T, Siclari F, Moccetti T, Vassalli G (2014) Extracellular vesicles from human cardiac progenitor cells inhibit cardiomyocyte apoptosis and improve cardiac function after myocardial infarction. Cardiovasc Res 103(4):530–541. doi:10.1093/cvr/cvu167

Boing AN, van der Pol E, Grootemaat AE, Coumans FA, Sturk A, Nieuwland R (2014) Single-step isolation of extracellular vesicles by size-exclusion chromatography. J Extracell Vesicles 3. doi:10.3402/jev.v3.23430

Braunwald E, Kloner RA (1985) Myocardial reperfusion: a double-edged sword? J Clin Invest 76(5):1713–1719

Cai ZP, Parajuli N, Zheng X, Becker L (2012) Remote ischemic preconditioning confers late protection against myocardial ischemia-reperfusion injury in mice by upregulating interleukin-10. Basic Res Cardiol 107(4):277. doi:10.1007/s00395-012-0277-1

Chen L, Wang Y, Pan Y, Zhang L, Shen C, Qin G, Ashraf M, Weintraub N, Ma G, Tang Y (2013) Cardiac progenitor-derived exosomes protect ischemic myocardium from acute ischemia/reperfusion injury. Biochem Biophys Res Commun. doi:10.1016/j.bbrc.2013.01.015

Cheng L, Sharples RA, Scicluna BJ, Hill AF (2014) Exosomes provide a protective and enriched source of miRNA for biomarker profiling compared to intracellular and cell-free blood. J Extracell Vesicles 3. doi:10.3402/jev.v3.23743

Chimenti I, Smith RR, Li TS, Gerstenblith G, Messina E, Giacomello A, Marban E (2010) Relative roles of direct regeneration versus paracrine effects of human cardiosphere-derived cells transplanted into infarcted mice. Circ Res 106(5):971–980. doi:10.1161/CIRCRESAHA.109.210682

Davidson SM, Selvaraj P, He D, Boi-Doku C, Yellon RL, Vicencio JM, Yellon DM (2013) Remote ischaemic preconditioning involves signalling through the SDF-1alpha/CXCR4 signalling axis. Basic Res Cardiol 108(5):377. doi:10.1007/s00395-013-0377-6

de Jong OG, Verhaar MC, Chen Y, Vader P, Gremmels H, Posthuma G, Schiffelers RM, Gucek M, van Balkom BWM (2012) Cellular stress conditions are reflected in the protein and RNA content of endothelial cell-derived exosomes. J Extracell Vesicles 1. doi:10.3402/jev.v1i0.18396

Del Conde I, Shrimpton CN, Thiagarajan P, Lopez JA (2005) Tissue-factor-bearing microvesicles arise from lipid rafts and fuse with activated platelets to initiate coagulation. Blood 106(5):1604–1611. doi:10.1182/blood-2004-03-1095

Deregibus MC, Cantaluppi V, Calogero R, Lo Iacono M, Tetta C, Biancone L, Bruno S, Bussolati B, Camussi G (2007) Endothelial progenitor cell derived microvesicles activate an angiogenic

program in endothelial cells by a horizontal transfer of mRNA. Blood 110(7):2440–2448. doi:10.1182/blood-2007-03-078709
Dragovic RA, Gardiner C, Brooks AS, Tannetta DS, Ferguson DJ, Hole P, Carr B, Redman CW, Harris AL, Dobson PJ, Harrison P, Sargent IL (2011) Sizing and phenotyping of cellular vesicles using nanoparticle tracking analysis. Nanomedicine 7(6):780–788. doi:10.1016/j.nano.2011.04.003
Forterre A, Jalabert A, Chikh K, Pesenti S, Euthine V, Granjon A, Errazuriz E, Lefai E, Vidal H, Rome S (2014) Myotube-derived exosomal miRNAs downregulate Sirtuin1 in myoblasts during muscle cell differentiation. Cell Cycle 13(1):78–89. doi:10.4161/cc.26808
Gaceb A, Martinez MC, Andriantsitohaina R (2014) Extracellular vesicles: new players in cardiovascular diseases. Int J Biochem Cell Biol 50:24–28. doi:10.1016/j.biocel.2014.01.018
Gatti S, Bruno S, Deregibus MC, Sordi A, Cantaluppi V, Tetta C, Camussi G (2011) Microvesicles derived from human adult mesenchymal stem cells protect against ischaemia-reperfusion-induced acute and chronic kidney injury. Nephrol Dial Transplant 26(5):1474–1483. doi:10.1093/ndt/gfr015
Genneback N, Hellman U, Malm L, Larsson G, Ronquist G, Waldenstrom A, Morner S (2013) Growth factor stimulation of cardiomyocytes induces changes in the transcriptional contents of secreted exosomes. J Extracell Vesicles 2. doi:10.3402/jev.v2i0.20167
Giricz Z, Varga ZV, Baranyai T, Sipos P, Paloczi K, Kittel A, Buzas E, Ferdinandy P (2014) Cardioprotection by remote ischemic preconditioning of the rat heart is mediated by extracellular vesicles. J Mol Cell Cardiol 68:75–78. doi:10.1016/j.yjmcc.2014.01.004
Gray WD, French KM, Ghosh-Choudhary S, Maxwell JT, Brown ME, Platt MO, Searles CD, Davis ME (2015) Identification of therapeutic covariant microRNA clusters in hypoxia-treated cardiac progenitor cell exosomes using systems biology. Circ Res 116(2):255–263. doi:10.1161/CIRCRESAHA.116.304360
Gupta S, Knowlton AA (2007) HSP60 trafficking in adult cardiac myocytes: role of the exosomal pathway. Am J Physiol Heart Circ Physiol 292(6):H3052–H3056. doi:10.1152/ajpheart.01355.2006
Hausenloy DJ, Yellon DM (2008) Remote ischaemic preconditioning: underlying mechanisms and clinical application. Cardiovasc Res 79(3):377–386. doi:10.1093/cvr/cvn114
Hausenloy DJ, Yellon DM (2009) Cardioprotective growth factors. Cardiovasc Res 83(2):179–194. doi:10.1093/cvr/cvp062
Hausenloy DJ, Yellon DM (2013) Myocardial ischemia-reperfusion injury: a neglected therapeutic target. J Clin Invest 123(1):92–100. doi:10.1172/JCI62874
Hausenloy DJ, Tsang A, Yellon DM (2005) The reperfusion injury salvage kinase pathway: a common target for both ischemic preconditioning and postconditioning. Trends Cardiovasc Med 15(2):69–75
Hausenloy DJ, Baxter G, Bell R, Botker HE, Davidson SM, Downey J, Heusch G, Kitakaze M, Lecour S, Mentzer R, Mocanu MM, Ovize M, Schulz R, Shannon R, Walker M, Walkinshaw G, Yellon DM (2010) Translating novel strategies for cardioprotection: the Hatter workshop recommendations. Basic Res Cardiol 105(6):677–686. doi:10.1007/s00395-010-0121-4
Heijnen HF, Schiel AE, Fijnheer R, Geuze HJ, Sixma JJ (1999) Activated platelets release two types of membrane vesicles: microvesicles by surface shedding and exosomes derived from exocytosis of multivesicular bodies and alpha-granules. Blood 94(11):3791–3799
Hergenreider E, Heydt S, Treguer K, Boettger T, Horrevoets AJ, Zeiher AM, Scheffer MP, Frangakis AS, Yin X, Mayr M, Braun T, Urbich C, Boon RA, Dimmeler S (2012) Atheroprotective communication between endothelial cells and smooth muscle cells through miRNAs. Nat Cell Biol 14(3):249–256. doi:10.1038/ncb2441
Hood JL, San RS, Wickline SA (2011) Exosomes released by melanoma cells prepare sentinel lymph nodes for tumor metastasis. Cancer Res 71(11):3792–3801. doi:10.1158/0008-5472.CAN-10-4455
Jansen F, Yang X, Proebsting S, Hoelscher M, Przybilla D, Baumann K, Schmitz T, Dolf A, Endl E, Franklin BS, Sinning JM, Vasa-Nicotera M, Nickenig G, Werner N (2014) MicroRNA expression in circulating microvesicles predicts cardiovascular events in patients with coronary artery disease. J Am Heart Assoc 3(6), e001249. doi:10.1161/JAHA.114.001249

Johnstone RM, Adam M, Hammond JR, Orr L, Turbide C (1987) Vesicle formation during reticulocyte maturation. Association of plasma membrane activities with released vesicles (exosomes). J Biol Chem 262(19):9412–9420

Kapustin AN, Chatrou ML, Drozdov I, Zheng Y, Davidson SM, Soong DY, Furmanik M, Sanchis P, de Rosales RT, Alvarez-Hernandez D, Shroff R, Yin X, Muller KH, Skepper JN, Mayr M, Reutelingsperger C, Chester A, Bertazzo S, Schurgers LJ, Shanahan CM (2015) Vascular smooth muscle cell calcification is mediated by regulated exosome secretion. Circ Res. doi:10.1161/CIRCRESAHA.116.305012

Kowal J, Tkach M, Thery C (2014) Biogenesis and secretion of exosomes. Curr Opin Cell Biol 29:116–125. doi:10.1016/j.ceb.2014.05.004

Lai RC, Arslan F, Lee MM, Sze NS, Choo A, Chen TS, Salto-Tellez M, Timmers L, Lee CN, El Oakley RM, Pasterkamp G, de Kleijn DP, Lim SK (2010a) Exosome secreted by MSC reduces myocardial ischemia/reperfusion injury. Stem Cell Res 4(3):214–222. doi:10.1016/j.scr.2009.12.003

Lai RC, Arslan F, Tan SS, Tan B, Choo A, Lee MM, Chen TS, Teh BJ, Eng JK, Sidik H, Tanavde V, Hwang WS, Lee CN, El Oakley RM, Pasterkamp G, de Kleijn DP, Tan KH, Lim SK (2010b) Derivation and characterization of human fetal MSCs: an alternative cell source for large-scale production of cardioprotective microparticles. J Mol Cell Cardiol 48(6):1215–1224. doi:10.1016/j.yjmcc.2009.12.021

Lonborg J, Vejlstrup N, Kelbaek H, Holmvang L, Jorgensen E, Helqvist S, Saunamaki K, Ahtarovski KA, Botker HE, Kim WY, Clemmensen P, Engstrom T (2013) Final infarct size measured by cardiovascular magnetic resonance in patients with ST elevation myocardial infarction predicts long-term clinical outcome: an observational study. Eur Heart J Cardiovasc Imaging 14(4):387–395. doi:10.1093/ehjci/jes271

Lotvall J, Hill AF, Hochberg F, Buzas EI, Di Vizio D, Gardiner C, Gho YS, Kurochkin IV, Mathivanan S, Quesenberry P, Sahoo S, Tahara H, Wauben MH, Witwer KW, Thery C (2014) Minimal experimental requirements for definition of extracellular vesicles and their functions: a position statement from the International Society for Extracellular Vesicles. J Extracell Vesicles 3:26913. doi:10.3402/jev.v3.26913

Mackie AR, Klyachko E, Thorne T, Schultz KM, Millay M, Ito A, Kamide CE, Liu T, Gupta R, Sahoo S, Misener S, Kishore R, Losordo DW (2012) Sonic hedgehog-modified human CD34+ cells preserve cardiac function after acute myocardial infarction. Circ Res 111(3):312–321. doi:10.1161/CIRCRESAHA.112.266015

Makridakis M, Roubelakis MG, Vlahou A (2013) Stem cells: insights into the secretome. Biochim Biophys Acta 1834(11):2380–2384. doi:10.1016/j.bbapap.2013.01.032

Mathivanan S, Ji H, Simpson RJ (2010) Exosomes: extracellular organelles important in intercellular communication. J Proteomics 73(10):1907–1920. doi:10.1016/j.jprot.2010.06.006

Mathur S, Walley KR, Wang Y, Indrambarya T, Boyd JH (2011) Extracellular heat shock protein 70 induces cardiomyocyte inflammation and contractile dysfunction via TLR2. Circ J 75(10):2445–2452

Messina E, De Angelis L, Frati G, Morrone S, Chimenti S, Fiordaliso F, Salio M, Battaglia M, Latronico MV, Coletta M, Vivarelli E, Frati L, Cossu G, Giacomello A (2004) Isolation and expansion of adult cardiac stem cells from human and murine heart. Circ Res 95(9):911–921. doi:10.1161/01.RES.0000147315.71699.51

Nolte-'t Hoen EN, van der Vlist EJ, Aalberts M, Mertens HC, Bosch BJ, Bartelink W, Mastrobattista E, van Gaal EV, Stoorvogel W, Arkesteijn GJ, Wauben MH (2012) Quantitative and qualitative flow cytometric analysis of nanosized cell-derived membrane vesicles. Nanomedicine 8(5):712–720. doi:10.1016/j.nano.2011.09.006

Ong SG, Lee WH, Huang M, Dey D, Kodo K, Sanchez-Freire V, Gold JD, Wu JC (2014) Cross talk of combined gene and cell therapy in ischemic heart disease: role of exosomal microRNA transfer. Circulation 130(11 Suppl 1):S60–S69. doi:10.1161/CIRCULATIONAHA.113.007917

Peche H, Heslan M, Usal C, Amigorena S, Cuturi MC (2003) Presentation of donor major histocompatibility complex antigens by bone marrow dendritic cell-derived exosomes modulates allograft rejection. Transplantation 76(10):1503–1510. doi:10.1097/01.TP.0000092494.75313.38

Piper HM, Garcia-Dorado D, Ovize M (1998) A fresh look at reperfusion injury. Cardiovasc Res 38(2):291–300

Raposo G, Stoorvogel W (2013) Extracellular vesicles: exosomes, microvesicles, and friends. J Cell Biol 200(4):373–383. doi:10.1083/jcb.201211138

Sahoo S, Klychko E, Thorne T, Misener S, Schultz KM, Millay M, Ito A, Liu T, Kamide C, Agrawal H, Perlman H, Qin G, Kishore R, Losordo DW (2011) Exosomes from human CD34(+) stem cells mediate their proangiogenic paracrine activity. Circ Res 109(7):724–728. doi:10.1161/CIRCRESAHA.111.253286

Scholz T, Temmler U, Krause S, Heptinstall S, Losche W (2002) Transfer of tissue factor from platelets to monocytes: role of platelet-derived microvesicles and CD62P. Thromb Haemost 88(6):1033–1038. doi:10.1267/th02121033

Sheldon H, Heikamp E, Turley H, Dragovic R, Thomas P, Oon CE, Leek R, Edelmann M, Kessler B, Sainson RC, Sargent I, Li JL, Harris AL (2010) New mechanism for Notch signaling to endothelium at a distance by Delta-like 4 incorporation into exosomes. Blood 116(13):2385–2394. doi:10.1182/blood-2009-08-239228

Shen B, Wu N, Yang JM, Gould SJ (2011) Protein targeting to exosomes/microvesicles by plasma membrane anchors. J Biol Chem 286(16):14383–14395. doi:10.1074/jbc.M110.208660

Skalnikova H, Motlik J, Gadher SJ, Kovarova H (2011) Mapping of the secretome of primary isolates of mammalian cells, stem cells and derived cell lines. Proteomics 11(4):691–708. doi:10.1002/pmic.201000402

Skog J, Wurdinger T, van Rijn S, Meijer DH, Gainche L, Sena-Esteves M, Curry WT Jr, Carter BS, Krichevsky AM, Breakefield XO (2008) Glioblastoma microvesicles transport RNA and proteins that promote tumour growth and provide diagnostic biomarkers. Nat Cell Biol 10(12):1470–1476. doi:10.1038/ncb1800

Smalley DM, Root KE, Cho H, Ross MM, Ley K (2007) Proteomic discovery of 21 proteins expressed in human plasma-derived but not platelet-derived microparticles. Thromb Haemost 97(1):67–80

Smith RR, Barile L, Cho HC, Leppo MK, Hare JM, Messina E, Giacomello A, Abraham MR, Marban E (2007) Regenerative potential of cardiosphere-derived cells expanded from percutaneous endomyocardial biopsy specimens. Circulation 115(7):896–908. doi:10.1161/CIRCULATIONAHA.106.655209

Sreekumar PG, Kannan R, Kitamura M, Spee C, Barron E, Ryan SJ, Hinton DR (2010) AlphaB crystallin is apically secreted within exosomes by polarized human retinal pigment epithelium and provides neuroprotection to adjacent cells. PLoS One 5(10), e12578. doi:10.1371/journal.pone.0012578

Thery C, Amigorena S, Raposo G, Clayton A (2006) Isolation and characterization of exosomes from cell culture supernatants and biological fluids. Curr Protoc Cell Biol Chapter 3:Unit 3.22. doi:10.1002/0471143030.cb0322s30

Timmers L, Lim SK, Arslan F, Armstrong JS, Hoefer IE, Doevendans PA, Piek JJ, El Oakley RM, Choo A, Lee CN, Pasterkamp G, de Kleijn DP (2007) Reduction of myocardial infarct size by human mesenchymal stem cell conditioned medium. Stem Cell Res 1(2):129–137. doi:10.1016/j.scr.2008.02.002

Valadi H, Ekstrom K, Bossios A, Sjostrand M, Lee JJ, Lotvall JO (2007) Exosome-mediated transfer of mRNAs and microRNAs is a novel mechanism of genetic exchange between cells. Nat Cell Biol 9(6):654–659. doi:10.1038/ncb1596

Van Deun J, Mestdagh P, Sormunen R, Cocquyt V, Vermaelen K, Vandesompele J, Bracke M, De Wever O, Hendrix A (2014) The impact of disparate isolation methods for extracellular vesicles on downstream RNA profiling. J Extracell Vesicles 3. doi:10.3402/jev.v3.24858

Vicencio JM, Yellon DM, Sivaraman V, Das D, Boi-Doku C, Arjun S, Zheng Y, Riquelme JA, Kearney J, Sharma V, Multhoff G, Hall AR, Davidson SM (2015) Plasma exosomes protect the myocardium from ischemia-reperfusion injury. J Am Coll Cardiol 65(15):1525–1536

Villarroya-Beltri C, Gutierrez-Vazquez C, Sanchez-Cabo F, Perez-Hernandez D, Vazquez J, Martin-Cofreces N, Martinez-Herrera DJ, Pascual-Montano A, Mittelbrunn M, Sanchez-Madrid F (2013) Sumoylated hnRNPA2B1 controls the sorting of miRNAs into exosomes through binding to specific motifs. Nat Commun 4:2980. doi:10.1038/ncomms3980

Waldenstrom A, Genneback N, Hellman U, Ronquist G (2012) Cardiomyocyte microvesicles contain DNA/RNA and convey biological messages to target cells. PLoS One 7(4), e34653. doi:10.1371/journal.pone.0034653

Wang X, Huang W, Liu G, Cai W, Millard RW, Wang Y, Chang J, Peng T, Fan GC (2014) Cardiomyocytes mediate anti-angiogenesis in type 2 diabetic rats through the exosomal transfer of miR-320 into endothelial cells. J Mol Cell Cardiol 74:139–150. doi:10.1016/j.yjmcc.2014.05.001

Welton JL, Webber JP, Botos LA, Jones M, Clayton A (2015) Ready-made chromatography columns for extracellular vesicle isolation from plasma. J Extracell Vesicles 4:27269. doi:10.3402/jev.v4.27269

Wolf P (1967) The nature and significance of platelet products in human plasma. Br J Haematol 13(3):269–288

Wolfers J, Lozier A, Raposo G, Regnault A, Thery C, Masurier C, Flament C, Pouzieux S, Faure F, Tursz T, Angevin E, Amigorena S, Zitvogel L (2001) Tumor-derived exosomes are a source of shared tumor rejection antigens for CTL cross-priming. Nat Med 7(3):297–303. doi:10.1038/85438

Xin H, Li Y, Cui Y, Yang JJ, Zhang ZG, Chopp M (2013) Systemic administration of exosomes released from mesenchymal stromal cells promote functional recovery and neurovascular plasticity after stroke in rats. J Cereb Blood Flow Metab 33(11):1711–1715. doi:10.1038/jcbfm.2013.152

Yellon DM, Davidson SM (2014) Exosomes: nanoparticles involved in cardioprotection? Circ Res 114(2):325–332. doi:10.1161/CIRCRESAHA.113.300636

Yu B, Kim HW, Gong M, Wang J, Millard RW, Wang Y, Ashraf M, Xu M (2014) Exosomes secreted from GATA-4 overexpressing mesenchymal stem cells serve as a reservoir of anti-apoptotic microRNAs for cardioprotection. Int J Cardiol 182C:349–360. doi:10.1016/j.ijcard.2014.12.043

Yuana Y, Bertina RM, Osanto S (2011) Pre-analytical and analytical issues in the analysis of blood microparticles. Thromb Haemost 105(3):396–408. doi:10.1160/TH10-09-0595

Yuana Y, Levels J, Grootemaat A, Sturk A, Nieuwland R (2014) Co-isolation of extracellular vesicles and high-density lipoproteins using density gradient ultracentrifugation. J Extracell Vesicles 3. doi:10.3402/jev.v3.23262

Zhang HG, Grizzle WE (2011) Exosomes and cancer: a newly described pathway of immune suppression. Clin Cancer Res 17(5):959–964. doi:10.1158/1078-0432.CCR-10-1489

Zhang Y, Zhang X, Shan P, Hunt CR, Pandita TK, Lee PJ (2013) A protective Hsp70-TLR4 pathway in lethal oxidant lung injury. J Immunol 191(3):1393–1403. doi:10.4049/jimmunol.1300052

Zhou Y, Xu H, Xu W, Wang B, Wu H, Tao Y, Zhang B, Wang M, Mao F, Yan Y, Gao S, Gu H, Zhu W, Qian H (2013) Exosomes released by human umbilical cord mesenchymal stem cells protect against cisplatin-induced renal oxidative stress and apoptosis in vivo and in vitro. Stem Cell Res Ther 4(2):34. doi:10.1186/scrt194

Part III
Stem Cells and Stem Cell-related Strategies for Heart Regeneration

Chapter 10
Delivery Modes for Cardiac Stem Cell Therapy

Neil Davies, Kyle Goetsch, Malebogo Ngoepe, Thomas Franz, and Sandrine Lecour

10.1 Introduction

Stem cell delivery to the heart is considered a potential breakthrough therapy for injury induced heart failure with particular reference to myocardial infarction. After a heart suffers a myocardial infarction, potentially at least a billion functioning cardiomyocytes can be lost due to the resultant ischaemia and this damage cannot be repaired in any meaningful manner by unaided endogenous repair mechanisms. Certainly this is the case in the human heart with the apparent very limited regenerative capacity of the human cardiomyocyte, with a renewal rate most reliably determined to be around 1–2 % per annum in the adult heart (Bergmann et al. 2009; Senyo et al. 2013). So after a myocardial infarction, the lost cardiomyocytes are inevitably replaced by a relatively avascular scar. This scar formation forms part of the complex and compensatory remodelling process, which also incorporates hypertrophy of the surviving cardiomyocytes, and allows for the heart to recover some function and strengthen the infarcted region. However though compensatory in the

N. Davies (✉) • K. Goetsch
Cardiovascular Research Unit, Department of Surgery, Faculty of Health Sciences, University of Cape Town, Observatory, South Africa
e-mail: neil.davies@uct.ac.za

M. Ngoepe
Department of Mechanical Engineering, University of Cape Town, Rondebosch, South Africa

T. Franz
Division of Biomedical Engineering, Department of Human Biology, Faculty of Health Sciences, University of Cape Town, Observatory, South Africa

S. Lecour
Cardioprotection Group, Hatter Institute for Cardiovascular Research in Africa and MRC Inter-University Cape Heart Group, Faculty of Health Sciences, University of Cape Town, Observatory, South Africa

R. Madonna (ed.), *Stem Cells and Cardiac Regeneration*, Stem Cell Biology and Regenerative Medicine, DOI 10.1007/978-3-319-25427-2_10

first instance, this process predisposes patients to a vicious and cyclical progression towards heart failure (Opie et al. 2006). Present optimal therapeutic interventions do not directly address the root cause of the heart failure, namely the death of cardiomyocytes and thus it is little surprise that stem cell based approaches with their potential promise of regeneration have received such wide spread attention over the last two decades. The vast majority of phase I/II clinical trials thus far have utilised the adult stem cell populations, in particular those derived from the bone marrow. This use of stem cell populations with ill-defined potential for differentiation toward cardiomyocyte lineage resulted in part from their ease of isolation, better safety profile, lack of ethical burden and initial small animal studies demonstrating striking cardiac tissue regeneration (Orlic et al. 2001). However, the hope of meaningful levels of regeneration due to transdifferentiation were quickly quashed by follow up animal studies (Murry et al. 2004) and also the, even when viewed optimistically, disappointing outcomes from the clinical trials. The most recent meta-analyses demonstrate this quite clearly with two meticulous analyses coming to differing conclusions, one finding modest improvements in a range of clinical end points (Fisher et al. 2015) and the other determining there to be no therapeutic benefit (Gyongyosi et al. 2015). It is also now widely accepted that when benefits are observed in these trials and related animal studies, they most likely predominately result from a paracrine effect (Gnecchi et al. 2008). There are many nuances in the relevant arguments with respect to efficacy of injected stem cells that are covered in detail elsewhere in this book but two threads that are common to reviews and analyses of adult stem cell delivery for myocardial infarction induced heart failure are the apparent safety of the approach and the very poor retention of delivered cells. The former finding is a necessity for any further work to proceed and the latter a conundrum for all cell-based approaches. It is beyond the scope of this chapter to consider the efficacy of stem cell delivery to the heart in restoring lost function and indeed this aspect is both hotly contested (see above) and covered by a plethora of meta-analyses. Most likely, several ongoing multi-center phase III trials will supply some of the definitive answers that the stem cell therapeutic community requires (de Jong et al. 2014). However, if the actual delivery of stem cells presently is inadequate and cannot achieve a therapeutic dosage, it is possible that these keenly awaited results may be less definitive in nature than hoped for. The future use of cells with greater potential for regeneration than bone marrow derived cells or the potency of any achieved paracrine effect will both be severely hampered by inefficient delivery. Thus in the first instance, we shall focus on studies, clinical and pre-clinical, where stem cells have been specifically tracked after being suitably labelled. These will be categorised according to their delivery routes.

10.2 Stem Cell Delivery Routes

In clinical trials, cells have been predominantly delivered to the heart either through infusion-based routes via major cardiac vessels or direct injection into the cardiac muscle (Fig. 10.1).

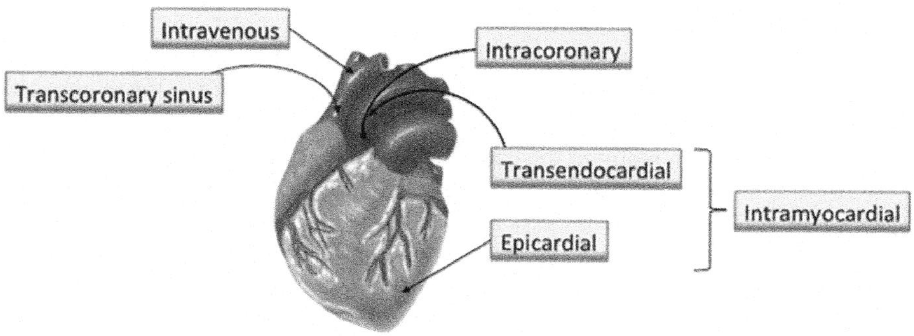

Fig. 10.1 Different delivery routes for cardiac cell delivery

10.2.1 Infusion Delivery

The leading route of delivery in clinical trials (Campbell and Suzuki 2012; Sanganalmath and Bolli 2013) has been through the infusion of a bolus of stem cells into the coronary vasculature with coronary arteries being the main vessels of choice. Systemic intravenous infusion has also been explored.

Intracoronary Infusion

In antegrade methods, the proximal aorta is reached usually by the femoral artery and the target coronary artery is then catheterised. In the clinic, cells are then normally infused under pressure to potentially facilitate ingress into the cardiac tissue. This is achieved by inflating a balloon and infusing the cells distally, thus limiting backflow of cells and also coronary flow forward past the balloon. It is envisaged that this scenario will assist cell delivery by increasing residence time in the infarcted region. Subsequently, several rounds (3–4) of pressurised cellular infusions are then applied. Certainly, this route of delivery is favoured as the access method is one most interventional cardiologists are familiar with, it can be achieved without complex mapping required for catheter based intramyocardial (IM) injections and has the potential to be used in high risk patients due to its minimally invasive nature. Further to this, it may have the potential to aid engraftment, as cells will necessarily tend to reach reasonably perfused cardiac tissue that might be less hostile to cell survival. Of course as a corollary to this, more compromised regions of the infarct may be poorly supplied. Concerns exist regarding the safety of the technique for the delivery of larger stem cells such as mesenchymal stem cells (MSC) with reports of pulmonary embolisms resulting from their migration away from the heart (Freyman et al. 2006; Furlani et al. 2009; Vulliet et al. 2004).

Retrograde delivery where access for the percutaneous catheter is gained through the femoral vein via the right atrium, cannulating the coronary sinus and catheterising the target coronary vein has received less attention than the coronary artery route. This method where cells are also infused under pressure avoids issues of occluded arteries precluding access but due the tortuous nature of the venous system is more technically challenging.

Systemic Intravenous Infusion

Intravenous injection has received quite significant interest due to the simple and non-invasive manner through which delivery might be achieved. Successful delivery would be reliant on homing to the cardiac site of injury and this has been demonstrated for cells such as bone marrow mononuclear cells (BMMNC) (Abbott et al. 2004). It is considered to most likely be effective in the context of acute myocardial infarction due to homing signals production being most pronounced at this stage (Frangogiannis 2006). However, it has been calculated based on blood flow rate to the left ventricle that infused cells would have to circulate many times to come into contact with the region of injury allowing time for deposition in other organs (Strauer and Steinhoff 2011).

10.2.2 Intramyocardial Injection

As implied, intramyocardial (IM) injection refers to the most direct way of delivering cells to the injured heart, namely by injection either epicardially, endocardially or through the coronaries. Epicardial delivery is mainly achieved through surgical exposure of the heart and the latter two routes through catheterisation.

Epicardial Delivery

Direct injection into the exposed heart has significant appeal due to the ease with which the injured or scarred region can be visualised and cells precisely delivered to infarcted and/or peri-infarct tissue. Multiple injections with large numbers of cells are also feasible and as with other IM routes, the need for reliance on homing of cells is removed. Due the reduced risk of coronary embolism, larger stem cells such as MSC can be delivered by this route. However, the invasive nature of this approach has generally limited its application clinically to studies where cell delivery is achieved in conjunction with open-heart surgeries such as coronary artery bypass grafting. This unavoidably reduces the number of patients who can be treated in this way. There have been limited reports on the use of minithoracotomies (Klein et al. 2007; Pompilio et al. 2008) to access the heart for direct injection but greater emphasis has been placed on development of catheter based IM injection methods.

Endocardial Delivery

Catheterisation is achieved usually through the aortic valve and injections are delivered perpendicularly to the endocardial surface. Placement is typically guided by electromechanical mapping using the NOGA® system (Biosense Webster, Diamond Bar, CA, USA) that allows for visualisation of the infarct and peri-infarct zones (Kharlamov et al. 2013). Thus this approach presents most of the advantages of direct surgical delivery in a minimally invasive manner and has been shown to be feasible and safe (Perin et al. 2003). However, the electromechanical mapping procedure is technically challenging and the equipment expensive.

Trans-coronary Delivery

The coronary sinus is accessed via the venous system with the catheter being passed though the right atrium. Injection is achieved parallel to the direction of the vein and it has been postulated that this may improve engraftment relative to the perpendicular manner in which cells are injected endocardially. The technique has been shown to be feasible and safe with BMMNC in pigs (George et al. 2008) and with skeletal myoblasts in a small clinical trial (Siminiak et al. 2005).

10.3 Engraftment

We have in the main focussed on studies that utilise in vivo imaging techniques to track cellular engraftment within the heart as these approaches allow for longitudinal analysis of cellular fate. They also avoid the potential inaccuracy of histological techniques were only a small fraction of the target area is visualised (Freyman et al. 2006).

The imaging techniques commonly used in cardiac studies (reviewed in (Azene et al. 2014; Chen and Wu 2013; Ransohoff and Wu 2010)) to quantify cellular engraftment exploit either direct physical labelling of the cell population or the introduction of a reporter gene to the cell. Direct labelling has thus far been the method of choice for clinical trials where cellular retention has been analysed, in part because it avoids the alteration of the genetic makeup of the delivered cells with its associated concern for possible mutagenesis. Cells labelled with a suitable contrast agent can then be imaged using the clinically available modalities of positron emission tomography (PET) and single-photon emission computed tomography (SPECT). PET and SPECT have been most widely used, usually with ^{18}F-FDG as a label for the former and ^{99m}TC-HMPAO the latter. PET scanners have a 1–2 order higher sensitivity than SPECT, detecting label at 10^{-11} to 10^{-12} mol/L. Cells that have been directly labelled are only really suitable for acute retention studies due to the labels having relatively short half-lives and concerns regarding retention of the label at the site after cell death through for example macrophage engulfment. The use of reporter genes is required for long term tracking as once their stable transfection or transduction into the cell's genome is achieved, cell signal should be proportional to the number of live cells. The Herpes Simplex Virus type 1 thymidine kinase (HSV1-TK) and thyroidal sodium iodide symporter (NIS) are suitable for PET and SPECT respectively when used in conjunction with imaging probes such as [18 F]FEAU (Perin et al. 2011) and ^{123}I (Templin et al. 2012). However, bioluminescence imaging (BLI) surpasses these techniques in sensitivity by at least 3 orders of magnitude (10^{-15} mol/L) through the detection of light generated from a reporter gene such as firefly luciferase. This modality at present only finds utility in small animal models due to attenuation of the light as it travels through tissue.

10.3.1 Infusion Studies

There have been a number of clinical studies where acute cellular retention has been assayed for infusion based delivery but not for IM (Table 10.1). The first PET feasibility study was carried out by Hofmann et al. (2005) to assess the retention of autologous BMMNC 75 min after either anterograde intracoronary (IC) or intravenous (IV) delivery to patients who had suffered a myocardial infarction 5–10 days prior. Only 1.3–2.6 % of the 18-F-FDG cells were retained in the heart with IC delivery and none could be detected after IV infusion. The majority of the cells were observed to have accumulated in the liver and spleen. A similar outcome was observed in a slightly larger study of similar design where 17 and 3 patients (3 to >300 days post-myocardial infarction) were enrolled for IC and IV delivery of peripheral blood stem cells (PBSC), respectively (Kang et al. 2006). Again IV delivery did not result in any detectable levels of cells in the heart and were only found in the lungs 2 h later. 2 % of IC delivered cells remained in the heart with the remainder lodged in the spleen, liver and bone marrow.

A variety of animal trials largely substantiate the observations of very poor levels of homing after IV delivery. In a canine study, 111In oxine-labeled bone marrow MSC (BMMSC) were found by SPECT to mainly reach the lungs after 24 h but there was a low level of redistribution to the heart (1.65 % of lung uptake) by 48 h indicating possible homing (Kraitchman et al. 2005). By 7 days as assessed by radioactive counting only 0.05 % of cells infused persisted in the heart. In a mouse BLI study where luciferase expressing BMMNC were injected through the tail vein into either sham operated or ischaemia/reperfusion (I/R) infarcted mice, a significant increase in homing to the infarcted mice's hearts was observed (Sheikh et al. 2007). However overall levels of engraftment were very low with only 0.1 % of cells residing in cardiac tissue at 14 days post-injection.

The majority of clinical and animal studies examining cell engraftment after IC delivery rendered similar results to the two IC/IV comparative studies described above. Interestingly, in two porcine studies retention of BMMNC was assessed after IC delivery with or without balloon occlusion and found to not significantly differ at either 1 h or 24 h (Doyle et al. 2007; Tossios et al. 2008). This is suggestive that the homing signals expressed by the injured heart are insufficient to enable cellular egress from the circulatory system in the short period of blocked blood flow before the cells are washed away by the released blood.

Initial comparison of the retrograde and more commonly utilised anterograde IC routes in a porcine model showed no significant difference in cellular engraftment at 1 h with 2.6 % and 3.2 % retention of BMMNC for IC and retrograde coronary venous (RCV) delivery (Hou et al. 2005). Two follow up studies indicate that perhaps IC is slightly more effective than RCV with initially higher levels of adipose

Table 10.1 Retention of cell delivery for cardiac repair

Cell type	Model	Route	Duration	Retention	Reference
BMMNC	Human	IC, IV	75 min	1.3–2.6 %—IC, no cell detection—IV	Hofmann et al. (2005)
PBSC	Human	IC, IV	2 h	2 %—IC, no cell detection—IV	Kang et al. (2006)
BMMSC	Canine	IV	7 days	0.05 %	Kraitchman et al. (2005)
BMMNC	Murine	IV	14 days	0.1 %	Sheikh et al. (2007)
BMMNC	Porcine	IC	1 h	8.7 %—balloon occlusion, 17.8 %—single-bolus	Doyle et al. (2007)
BMMNC	Porcine	IC, IM	24 h	3.3 %—IC (balloon), 3 %—IC (without), 15 %—IM	Tossios et al. (2008)
BMMNC	Porcine	IC, IM, RCV	1 h	2.6 %—IC, 11 %—IM, 3.2 %—RCV	Hou et al. (2005)
AdMSC	Porcine	IC, RCV	1 h and 24 h	1 h—57.2 %—IC, 17.9 %—RCV 24 h—22.6 %—IC, 18.7 %—RCV	Hong et al. (2014b)
BMMNC	Porcine	IC, IM	1 h	13,579 cells—IM, 7049 cell—IC	George et al. (2008)
BMMSC	Rat	IV, IM	1 week	15 %—IM, no cell retention—IV	Hale et al. (2008)
BMMNC	Rat	IV, LV, IM	24 h	16 %—IM, <1 %—IV, <1 %—LV	Nakamuta et al. (2009)
iPSC-CM	Murine	IM	4 weeks	8 %	Lepperhof et al. (2014)
hCPC	Murine	IM	4 weeks	<5 %	Liu et al. (2012a)
CSC	Murine	IC	24 h	12.7 %	Hong et al. (2014a)
BMMSC	Porcine	IM	10 days	5.8 %	Gyongyosi et al. (2008)
BMMSC	Porcine	EC	35 days	40–50 %	Perin et al. (2011)
iPSC	Porcine	IM	15 weeks	≈2 %	Templin et al. (2012)
EPC	Canine	EC, IM	40 min	57 %—EC, 54 %—IM	Mitchell et al. (2010)

IC Intracoronary, *IV* Intravenous, *IM* Intramyocardial, *RCV* Retrograde coronary venous, *IA* intra-aortic, *LV* Left ventricular cavity, *EC* Endocardial
IPSC-CM induced pluripotent stem cell-derived cardiomyocytes, *hCPC* human cardiac progenitor cells, *CSC* cardiac stem cells, *EPC* Endothelial progenitor cells

derived MSC (AdMSC) engraftment in a porcine model of IC delivery at 24 h though this difference abated at 48 h (Hong et al. 2014b). In a clinical study (Silva et al. 2009), elevated levels of engraftment of ^{99m}TC-BMC were found with IC delivery at 4 and 24 h post administration.

In the majority of the IC, RCV and IV studies, cells were found preferentially located within the pulmonary system. This is of interest as it has been shown that deposition of human MSC into NOD/SCID mice lungs after IV infusion upregulated the anti-inflammatory protein TSG-6 that was then demonstrated to be responsible for a paracrine based reduction in infarct size (Lee et al. 2009).

10.3.2 Intramyocardial Injection

As noted above, thus far no clinical studies have been reported that assay retention and engraftment after intramyocardial (IM) injection. In one of the earlier studies in a porcine model, it was shown that acute retention at 1 h is significantly higher than the infusion methods (IM 11 %, IC 2.6 % and RCV 3.2 %) (Hou et al. 2005). In almost all other direct comparisons between IM and infusion based deliveries, this has been found to be the case (George et al. 2008; Hale et al. 2008; Li et al. 2009, 2011; Nakamuta et al. 2009; Tossios et al. 2008). It is perhaps not entirely surprising as the need for reliance on homing signals to attract cells out of the circulatory system is avoided. However levels achieved though higher than those by infusion are still low and reporter gene studies tend to indicate further rapid loss of cells from the heart. In BLI studies with a range of cell types, 90 % or more of cells that were initially retained are lost in the medium term (7–28 days post-injection) (Lepperhof et al. 2014; Li et al. 2011; Liu et al. 2012a; Westrich et al. 2010). In addition to these BLI studies, similar decreases in cell numbers in the hear were discerned using a novel real-time PCR analysis of infarcted mouse hearts injected with cardiac derived stem cells (CSC) (Hong et al. 2014a). When autologous porcine BMMSC expressing HSV1-TK and red fluorescent protein (RFP) were injected endocardially via catheterization under NOGA guidance and tracked by PET, signal was detected at 30 h post-injection but the signal was lost at 7 days (Gyongyosi et al. 2008). Though histological analysis for RFP expressing cells indicated around 6 % were still present at 7 days, this is similar to the trends observed in small animals. Very differently in a porcine study utilising HSV1-TK expressing porcine BMMSC where 20-fold greater cell numbers were delivered by NOGA guided catheter injection endocardially, an estimated 40–50 % of cells were seen to survive to 35 days in 3 pigs and in one pig to 5 months though with a decreased signal (Chen and Wu 2013; Perin et al. 2011). Interestingly in a porcine study where NIS expressing human iPSC (immunosuppression with cyclosporine A) were injected under NOGA guidance in similarly large numbers either with or without human MSC, only in the pigs that received iPSC injected with MSCs was a signal still detected by SPECT at 15 weeks (Templin et al. 2012). This raises the questions of whether very large numbers of MSC are required to create a more conducive environment for engraftment and whether this is a species-specific effect.

From the above it seems reasonable to assume that endocardial delivery via a catheter would be similar in cellular retention to that achieved with direct epicardial injection. A study in a canine model whereby 111In-tropolone labelled endothelial progenitor cells were either directly injected into the epicardium or endocardial injections were performed using the Stiletto Endo-myocardial Injection System

(Boston Scientific) under radio-graphic fluoroscopic guidance (Mitchell et al. 2010), equivalent levels of acute retention at 30–40 min post-injection were observed. It would seem that a retention study on IM delivery retention should be performed in humans to ascertain whether the higher, though still underwhelming, levels of cells retained with this route in animals are also seen in the clinic.

10.4 Aspects Influencing Cell Retention and Engraftment

Simple mechanical ejection from the delivery site could of course play a role in the immediate loss of cells. Intramyocardial injections of 18-F-FDG labelled CSC into infarcted rat hearts that had either been arrested or had their ventricular rate slowed with adenosine when visualized with PET at 1 h showed substantially improved retention for both conditions (control: 17.8 %; arrested: 75.6 %; adenosine: 35.4 %) (Terrovitis et al. 2009). A similar retention to that achieved with adenosine was seen when the injection hole was simply sealed with fibrin glue. The authors argue that the greater retention seen in the arrested heart may reflect that not only was potential mechanical ejection reduced in this condition but also washout by myocardial perfusion. It might be expected that the mechanical influence on retention would be more pronounced in a heart beating at 300–400 bpm than in the clinic but this has not been empirically determined yet.

An important aspect that might be expected to influence retention and long-term survival would be timing of delivery after infarction. The environment within the infarcted tissue changes dramatically during the remodelling process for many parameters that might influence cells inclusive of inflammation, ischaemia, extracellular matrix structure and biomechanics (Holmes et al. 2005). It is therefore perhaps somewhat surprising that in the relatively limited number of studies directly examining the consequence of timing, no marked effect was observed. In a human study where retention of 111IN-oxine labelled PBSC was assessed by SPECT, 6.3 % acute retention was observed in infarcts less than 14 days old relative to the 4.5 % in older infarcts (Schachinger et al. 2008). Small animal studies have seen similar results with little or no difference in retention observed whether cells are injected acutely or at later time points (Bonios et al. 2011; Nakamuta et al. 2009; Swijnenburg et al. 2010).

In a recent meta-analysis of stem cell therapy outcome in the clinic, a positive correlation between mononuclear cell dose infused and increase in ejection fraction was detected (Clifford et al. 2012). The limited studies that have assessed the impact of dosage on retention and survival have also found a similarly positive relationship. 1–2 % of either 10^5 or 10^6 BMMSC or BMMNC were found to survive at 6 weeks after injection into rat hearts (Muller-Ehmsen et al. 2006) and acute retention (Shen et al. 2012) was found to be around 10 % for all dosages (10^4 to 5×10^5) of cardiospheres (a natural mixture of resident cardiac stem cells and supporting cell types (Smith et al. 2007)) with a positive correlation between dosage and functional recovery. An elegant study by Liu et al (Liu et al. 2012a) took advantage of the apparent variability present in epicardial intramyocardial delivery (Hou et al. 2005) to stratify their treatment cohort (HSV1-TK expressing human cardiac progenitor cell (hCPC) delivered into infarcted SCID mice hearts) into high and low early engraftment groups. A clear and significant improvement in various left ventricular (LV) functional parameters was observed for the high engraftment group, which was ascribed to the paracrine effect.

These latter two studies emphasize that improving retention and survival of stem cells delivered will correspond to improved outcome. We shall devote the remainder of this chapter to exploring the utilization of injectable biomaterial scaffolds to achieve this desired improvement. Biomaterials are attractive as they in their various guises afford the possibility of tackling multiple factors that might influence engraftment such as mechanical entrapment and reduction of anoikis, inflammation and ischaemia.

10.5 Injectable Biomaterial Cellular Vehicles

10.5.1 Biomaterial Injection

Clearly, to act as cellular delivery vehicles injectable biomaterial scaffolds should gel sufficiently quickly after injection to effectively entrap cells, they must be biocompatible and allow for cellular adhesion to reduce anoikis. The ability to stimulate potentially cellular protective mechanisms such as angiogenesis would of course be desirable. These and other related aspects are the subjects of intensive recent research as will be seen below. However, injectable materials can be intrinsically cardioprotective. The cyclical process of pathological left ventricular dilation that ensues after a myocardial infarction is driven by increased stress in the wall (Opie et al. 2006; Sutton and Sharpe 2000). This increased stress can be considered to derive from the interaction between the increasing ventricular volume and the thinning of the infarcted wall as described by the Law of Laplace $T = \frac{p \cdot r}{h}$, where T is the tension in the cardiac wall, p is the blood pressure in the ventricular cavity, r is the radius of the ventricular cavity and h is the thickness of the cardiac wall. As can be seen from this, an increase in wall thickness will reduce the stress experienced by the cellular components of the ventricular wall and thus potentially inhibit the progression towards heart failure. Injection of a material within the wall can achieve such thickening and finite element models (Wall et al. 2006; Wenk et al. 2009) have shown that this resultant bulking can reduce stress by up to 20 % in the critical border zone region and slightly increase the ejection fraction (EF).

The above finite element modelling findings have been widely supported by studies that have looked at injection of a broad range of biomaterials into infarcted hearts, inclusive of both natural materials (e.g fibrin; collagen; matrigel; extracellular matrix derivatives; alginate) and synthetic (e.g. self-assembling peptides; polyethylene glycol; poly(N-isopropylacrlamide) polymers) (reviewed in Nelson et al. 2011, Radisic and Christman 2013). These studies have on the whole observed wall thickening and varying degrees of functional preservation. There is a paucity of studies that have empirically investigated the mechanism through which these above results were obtained. However, a recent study (Ifkovits et al. 2010) demonstrated as predicted by the mathematical models, a stiffer methacrylated hyaluronic acid hydrogel resulted in less infarct expansion and left ventricular dilation in a porcine infarction model. This type of research is urgently needed to guide the design and optimisation of injectable hydrogels for cardiac therapy. It should be noted that the above finding also raises potential further complexity in development of these hydrogels as cellular vehicles because the stiffness of hydrogels can significantly influence the behaviour of stem cells entrapped within them—both with respect to their ability to migrate within the hydrogel (Ehrbar et al. 2011) and their potential direction of differentiation (Pek et al. 2010).

As a stand-alone cardiac therapy, alginate has been the most intensively investigated. Alginate, an anionic polysaccharide derived from brown seaweed is biocompatible and widely used in the pharmaceutical and medical device industries. A low viscosity version of alginate has been developed that is injectable and polymerises spontaneously within the heart due to increasing calcium ion concentration in the infarct. Landa et al. (2008) investigated the effects of alginate delivery via epicardial injection into rat hearts 7 and 60 days after infarct induction. The alginate was gradually replaced by connective tissue over 6 weeks demonstrating biodegradability. It should be noted that though a non-degradable implant might be considered desirable as stress reduction might also be maintained, we and others have shown this to not be the case even when using very biocompatible polyethylene glycol (PEG) hydrogels (Dobner et al. 2009; Rane et al. 2011). Eight weeks after injection of alginate, infarct scar thickness was increased and left ventricle dilation reduced for injections at both time points though these improvements were diminished in the group receiving injections into the chronic infarct. The latter result emphasising the greater difficulty treating infarcted hearts at that late stage. Of interest, the positive outcomes in the 7 day injection group were at least comparable to a group that received 1×10^6 neonatal cardiomyocytes in saline. In a follow up large animal study, the alginate solution was delivered to infarcted porcine hearts at 4 days post-infarction and again reductions in ventricular dilation and increased scar thickness were observed (Leor et al. 2009). Interestingly, intracoronary delivery was realized through catheter injection within the left anterior descending artery and ingress into the infarcted tissue was achieved through the leaky vessels present in the infarct. Occlusion of the coronary was avoided, as calcium levels were only high enough within the tissue to achieve phase inversion to a hydrogel. This relatively simple minimally invasive approach is certainly appealing for clinical application of alginate as a stand-alone therapy but with respect to use as a route for enhancing cellular engraftment it may be less effective. The probable inefficient migration of cells from the vasculature into the myocardium that blights intracoronary delivery is likely not to be improved by delivery with a hydrogel.

These promising pre-clinical results with alginate resulted in an initial feasibility and safety trial in humans (ClinicalTrials.gov Identifier NCT00557531) whereby 2 mL of the alginate solution (termed IK-5001) was delivered via the infarcted coronary to 27 patients that had undergone a moderate to large myocardial infarction 7 days prior. The patients had been successfully revascularised. The treatment was well tolerated and preservation of LV indices was observed (Frey et al. 2014). This positive outcome has resulted in the enrolment of 300 patients with acute myocardial infarction into an ongoing multicentre, randomized and double-blind phase 2 clinical trial (ClinicalTrials.gov Identifier NCT01226563). In a related human study, an alternative form of alginate was delivered by epicardial intramyocardial injection in 10–15 sites in 11 dilated cardiomyopathy patients that were undergoing coronary artery bypass grafting (CABG) (ClinicalTrials.gov Identifier NCT00847964). In a small subset that could undergo MRI, it was determined by mathematical modelling that myofibre stress was reduced by 35 % (Lee et al. 2013). So though this outcome derives from only three patients and is complicated by the simultaneous CABG procedure, it is the first indication that hydrogel based stress reduction is achievable in humans.

The progression of these types of therapies to the clinic will be facilitated by their ability to be delivered by catheterization. This is a demanding goal as the polymer solutions need to have low enough viscosity to flow through the catheter and remain ungelled till reaching their target site upon which gelation must occur as rapidly as possible. Apart from alginate (see above), there are very few reports describing catheter delivery of a hydrogel to the heart. Recently though a hydrogel solution derived from ventricular extracellular matrix was successful injected endocardially into porcine hearts (Singelyn et al. 2012) using a NOGA® guided Myostar® catheter. In follow up study, the catheter delivered extracellular matrix hydrogel was shown to improve cardiac function in a porcine myocardial infarct model (Seif-Naraghi et al. 2013). More recently a pH-switchable hydrogel (ureido-pyrimidinone-modified PEG hydrogel) was used to deliver growth factors to infarcted porcine hearts (Bastings et al. 2014). Though probably not useful for cellular delivery due to the switch occurring as the solution at pH 8.5 transits to neutral in the heart, the type of rapid and controllable gelling achieved will be desirable for cellular vehicles.

These types of outcomes with biomaterial delivery alone have resulted in significant recent interest in determining the influence of co-delivery of injectable hydrogels with stem cells on both stem cell retention and efficacy.

10.5.2 Biomaterial-Cell Delivery

Biological Materials

Biological materials are inherently attractive as cellular vehicles as they usually have a good level of biocompatibility. They are also readily available but have batch-to-batch variability that can be eliminated in synthetic hydrogels.

Fibrin, a well characterised hydrogel that is derived from the mixing of a solution of fibrinogen with one containing thrombin, factor XIIIa and calcium, was one of the earliest materials used to deliver cells to the heart (Table 10.2). In the initial

Table 10.2 Cell delivery within biomaterials for cardiac repair

BMMSC	Rat	Collagen	1 week	23 % (cells), 16 % (collagen and cells)	No function improvement (collagen and cells)	Dai et al. (2009)
BMMNC	Rat	Hyaluronic acid	4 weeks	2.5-fold increase (cells and biomaterial)	EF: significant improvement with cell and biomaterial combination	Chen et al. (2013)
CDC	Murine	Hyaluronic acid/gelatin	24 h	35 % (cells and biomaterial)	Improved cardiac function (LVEF) $p<0.05$—cells and biomaterial	Cheng et al. (2012)
			21 days	6 % (cells and biomaterial)		
AdMSC	Rat	Chitosan	4 weeks	3.05 % (Cells and biomaterial)	LVEF: 59.29 % (cells and biomaterial), 52.92 %, (biomaterial), 49. 53 % (cells)	Liu et al. (2012b)
BMMSC	Porcine	Alginate	4 weeks	≈12 % (cells and biomaterial)	ND	Panda et al. (2014)
BMMSC	Rabbit	α-cyclodextrin/MPEG-PCL-MPEG	4 weeks	2.5-fold increase in engraftment (cells and biomaterial)	LVEF: 62.35 % (cells and biomaterial), 47.14 % (cells), 35.14 % (saline)	Wang et al. (2009)
ESC	Rat	Oligo[poly(ethylene glycol) fumarate]	1 day	1 day: 6.54 % (cells), 16.35 % (cells and biomaterial) 4 weeks: 4.56 % (cells), 11.83 % (cells and biomaterial)	FS: cells and biomaterial significantly ($p<0.01$) improved LV function compared to cells and biomaterial alone	Wang et al. (2012)
			4 weeks			
AdMSC	Rat	PNIPAAm	4 weeks	1.5-fold increase in engraftment (cells and biomaterial)	EF and FS: Significant improvement for cells and biomaterial combined compared to controls	Li et al. (2014)
BMMSC	Rat	Self-assembling peptides	4 weeks	Twofold increase in engraftment (cells and biomaterial)	EF: 59.31 % (cells and biomaterial), 48.31 % (cells), 42.06 % (saline)	Cui et al. (2010)
BMMSC	Rat	Self-assembling peptides	4 weeks	Fourfold increase in engraftment (cells and biomaterial)	EF: 53.06 % (cells and biomaterial), 31.23 % (biomaterial), 35.21 % (cells)	Guo et al. (2010)
BMMSC	Porcine	Self-assembling peptides	2 months	Fourfold increase in engraftment at 7 days (cells and biomaterial)	EF of cells delivered at day 1 post MI: 55.1 % (cells and biomaterial), 46.5 % (biomaterial), 46.3 % (cells)	Chang et al. (2015)
BMMSC	Porcine	Self-assembling peptides	4 weeks	Eightfold increase in engraftment at 7 days (cells and biomaterial)	Cells and biomaterial improved both diastolic and systolic function	Lin et al. (2010)

SkMC Skeletal myoblasts, *CDC* Cardiosphere-derived cells, *ESC* Embryonic stem cells, *ND* Not done studies, skeletal myoblasts were delivered with fibrin to infarcted rat hearts. The combination and controls were injected 1 week post-infarction and hearts were assessed by echocardiography and histology 4 weeks later (Christman et al. 2004a, b). As assessed by skeletal myosin immunohistochemistry, engraftment was similar for the cell and cell plus fibrin groups at 24 h but by 4 weeks post-injection a twofold increased engraftment of skeletal myoblasts was seen in cells plus fibrin. Similar functional improvements were seen for fibrin, cells and cells plus fibrin and all 3 groups were seen to elicit an angiogenic response. Thus in this instance the simpler route of only delivering the material without the skeletal myoblasts would appear optimal. In another fibrin based study, 99mTc-labelled BMDC were quantified 24 h after injection into infarcted rat hearts through counting the radioactivity in the excised organ in a gamma counter (Nakamuta et al. 2009). In this study, using a more global means of engraftment assessment, a 2.5-fold increase in cell retention was seen for the fibrin plus cells group. Again at 4 weeks post-infarction for both the cell alone and cell plus fibrin groups similar functional improvements were observed though with a trend towards greater scar thickness in the latter. In a BLI based analysis of retention of luciferase expressing AdMSC with or without fibrin injected into infarcted rat hearts, an increasing divergence between cells alone and cells with fibrin was quantified (1 day: 1.3 ×; 7 days: 3 ×, 14 days: 5 ×) till finally at day 28, a signal could readily be detected for fibrin plus cells but none for cells alone (Yang et al. 2013). It should be noted that though there was greater engraftment for fibrin plus cells, there was still a rapid decrease in signal with only about 2 % of the original 5 million AdMSC present in the heart. Finally with respect to fibrin, in a follow up to the study of Nakamuta et al. (2009), the functional recovery after AdMSC delivery to infarcted hearts was assessed in much greater detail (Danoviz et al. 2010). Stroke work measured in the presence of phenylephrine, a global index of cardiac function that depends on both pressure generation and ejection capability was seen to be improved by all cell containing groups but only returned to normality when hearts were injected with a biomaterial plus cells. There was a positive correlation between greater cellular engraftment ((AdMSC/media: 4.8 %; AdMSC/fibrin: 13.8 %; AdMSC/collagen 26.8 %) measured by 99 m-Tc labeling) and stroke work improvement.

In another study in a rat infarction model, collagen was used as a cellular vehicle for BMMSC labeled with europium nanoparticles (Dai et al. 2009). Engraftment was measured at 4 weeks post-infarction and no increase in engraftment was observed for the collagen plus cell group relative to cells delivered in saline unlike that observed in the study of Danoviz et al. (2010) at 24 h. However there was reduced traffic of cells to remote organs in the collagen plus cells group. Somewhat unexpectedly, and again in contrast to the Danoviz et al. (2010) study, the cells and collagen alone groups showed functional improvement relative to the control but the combined group failed to improve function. The reason for this difference is not clear though it is notable that the cells were delivered 24 h and 1 week post-infarction in the Danoviz et al. (2010) and Dai et al. (2009) studies respectively. We have observed markedly different distributions of polymers injected at either 1 week or shortly after infarction (Kadner et al. 2012). Biopolymers injected at 1 week assume a bulky monodisperse structure whilst those closer to point of infarction have a more delicate fibrillar structure and thus less isolated in distance from the surrounding tissue. Therefore diffusion of oxygen and nutrients to the cells might have been more effectively impeded when injected at 1 week within a long lasting biopolymer implant (Dai et al. 2009). Of course, other differences exist between these studies such as collagen source and cell type utilized that may have played a role in the different outcomes.

Hyaluronic acid is a nonsulfated glycosaminoglycan and component of the extracellular matrix that has been used to deliver BMMNC to infarcted rat hearts immediately post-infarction (Chen et al. 2013). Four weeks after delivery 2.5-fold higher levels of DiI labeled BMMNC were found in hearts where hyaluronic acid and cells were injected together than when BMMNC were delivered alone. Both single treatments reduced apoptosis and scar length and increased the ejection fraction (EF), angiogenesis and arteriogenesis but in all instances a further significant increase was observed for the combination. Very similar findings were reported for human cardiosphere-derived cells injected into the infarcted hearts of SCID mice when a hyaluronic acid/gelatin hybrid hydrogel was used as a delivery vehicle (Cheng et al. 2012). Even greater increases in retention were seen for this combination, with approximately 6–7 fold increases at 24 h and 21 days relative to cells alone. When a hyaluronic acid hydrogel without gelatin was used, no increase was observed at 24 h but the hyaluronic acid alone group was not utilized for the longer-term studies making it difficult to compare these two studies directly. Because overall engraftment in the biopolymer group reduced from 35 % at 24 h to 6 % at 21 days, the authors postulated that the improvements observed for apoptosis, angiogenesis, EF and viable tissue within the infarct were due to the paracrine effect though evidence of differentiation towards cardiomyocytes was observed.

In contrast to the above biological materials that are inherent to the mammalian extracellular matrix, others have investigated the utility of polysaccharides derived from sources such as crustacean shells and seaweed. In a series of studies, the ability of chitosan to improve cellular retention and therapeutic efficacy was assessed. A temperature responsive chitosan derivative (Chenite et al. 2000) that gels rapidly and spontaneously when the environmental temperature is increased was found to increase both retention and efficacy of embryonic stem cells (ESC) (Lu et al. 2009), ESC obtained from somatic cell nuclear transfer (Lu et al. 2010) and most recently AdMSC (Liu et al. 2012b) after injection into infarcted rat hearts. Similar results were found for all cell types but the retention of AdMSC was assessed globally by BLI after lentiviral transduction of the luciferase gene in contrast to histological evaluation used in the earlier studies. Relative to the PBS/AdMSC control group, retention of AdMSC in the chitosan group increased from 1.5-fold at day 1 to approximately eightfold at day 28 though even for this group overall cellular retention decreased to around 10 % of cellular retention detected at day 1. This improved retention was associated with functional recoveries in parameters such as EF and +dp/dt, decrease in apoptosis and increased wall thickness and neovascularization. Of note, evidence that chitosan scavenged reactive oxygen species was given as a possible contributing source for the positive outcome. Alginate, another polysaccharide, has been used as a vehicle for epicardial injection of BMMSC into the border zones of reperfused pig hearts 4 weeks after infarction induction (Panda et al. 2014). The alginate was covalently modified with the cellular adhesion peptide RGD, presumably to overcome the poor cellular adhesion observed for alginate. Interestingly, chitosan has similarly poor cellular adhesion (Luna et al. 2011) but no modifications were utilized in the studies detailed above. Possibly targeting cellular adhesion may further improve the engraftment of stem cells injected in conjunction with chitosan. Alginate RGD was shown to improve the retention of cells 2 weeks post delivery by fourfold but only when 2 % w/v solution was used and not with a 1 % w/v formulation. The effect of improved engraftment on function was not determined.

Thus, marked improvements in retention and therapeutic outcome have been reported when cells are delivered in conjunction with biological hydrogels. However, as indicated above concerns exist with respect to the batch-to-batch variability of these materials and therefore others have investigated synthetic hydrogels that allow for a more clinically acceptable standardisation of the cellular vehicle.

Synthetic Materials

Taking advantage of the rapid gelling of triblock polymer hydrogel (α-cyclodextrin/poly(ethylene glycol)–b-polycaprolactone-(dodecanedioic acid)-polycaprolactone–poly(ethylene glycol)), BMMSC were injected 7 days post infarction in a rabbit model (Wang et al. 2009). Histological evaluation of DAPI labeled cells showed a 2.5-fold increase in engraftment 4 weeks after delivery. This resulted in an elevated EF, decreased infarct size and increased neovascularization relative to the cells alone control. In a preceding study, the hydrogel was found to improve EF and reduce infarct size but did not influence vessel density suggesting that the latter effect resulted from the engrafted cells (Jiang et al. 2009). Another polyethylene glycol composite (oligo[poly(ethylene glycol) fumarate]) that polymerized in the presence of ammonium persulfate and tetramethylethylenediamine was used to deliver GFP labeled mouse ESC 1 week after rat hearts were infarcted (Wang et al. 2012). Co-delivery with the hydrogel resulted in a 2.5-fold increase in cellular retention and the hydrogel/cell combination augmented cardiac function, decreased infarct size and increased capillary density to a greater extent than that observed for single treatment controls. Interestingly, no teratoma formation was observed in this study.

Recently, brown fat AdMSC were injected with thermoresponsive poly (N-isopropylacrylamide) (PNIPAAm) hydrogels into rat hearts immediately after infarction (Li et al. 2014). The hydrogels were loaded with single-wall carbon nanotubes (SWCNT) to elicit improved cellular adhesion, spreading and proliferation within the PNIPAAm environment. Though enhancement of these parameters was observed in vitro, only a relatively modest increase in retention of about 1.5-fold for the combination of hydrogel and cells was seen at 4 weeks. However, significant improvements in cardiac function were observed for the PNIPAAm/AdMSC group, as too was decreased infarct size and increased wall thickness.

Injectable scaffolds can be formed from self-assembling peptides (SAP) and the sequence RARADADARARADADA-CNH2 (RAD16-II) has received particular attention as a cellular delivery vehicle. BMMSC that had been selected for positive expression of c-kit and the cardiac transcription factor NKx2.5 were delivered with RAD16-II 30 min after infarction induction in female rats (Cui et al. 2010). Cellular retention was determined by fluorescence in situ hybridization for the Y-chromosome of the injected male cells. A twofold increase in retention was observed at 4 weeks post-injection for cells plus nanofibers and again this increased retention was associated with further significant improvements to that gained with delivery of cells alone. Decreased infarct size, increased neovascularization and improved EF% resulted from the combined injection group. In a follow-up study (Guo et al. 2010), the SAP was synthesized with incorporation of the potent cell adhesion peptide RGDSP. This modification further increased the twofold retention seen with the unmodified RAD16-II to fourfold and further advances in infarct size, neovascularization and function were achieved.

Importantly the RAD16-II nanofiber scaffolds have been used to deliver cells in the porcine model. In sequential studies the utility of these scaffolds to influence the therapeutic outcome of freshly isolated BMMNC was assessed (Chang et al. 2015; Lin et al. 2010). In the initial investigation, DiI labeled BMMNC with RAD16-II were injected epicardially, immediately after infarction, into the infarct zone of mini-pig hearts (Lin et al. 2010). Scar length ratio% was decreased and scar thickness ratio% increased by all groups but the latter parameter was significantly increased by the combination relative to the other groups. Interestingly, injection of RAD16-II alone was found to predominately improve diastolic function and BMMNC alone mainly enhanced systolic function. Combining the two treatments resulted in an increase in both systolic and diastolic function. This therapeutic outcome was associated with an approximately eightfold increase in labeled cells detected histologically 4 weeks after infarction. Subsequently, the effect of timing of both delivery and isolation of BMMNC was assessed in the mini-pig model (Chang et al. 2015). Autologous BMMNC were isolated 1, 4 and 7 days after infarction and injected on the same day. Only cells isolated on day 1 had similar results to those observed in the immediate delivery study (see above) with cells at 4 days showing lesser therapeutic impact. BMMNC obtained 7 days post-infarction did not result in any improvement for the parameters measured. The proportion of CD14/CD16 positive cells in the BMMNC isolates decreased with time and it was suggested that this decrease in a potentially more cardioprotective sub-population might have contributed to the reduced efficacy. This result further amplifies the general requirement for much greater clarity with respect to these types of finer details in the cell therapy.

Growth Factor Delivery Systems

It is naturally appealing to look to combine the delivery of stem cells with a biomaterial that releases growth factors. The controlled release of growth factors can not only be inherently cardioprotective but also cytoprotective towards the delivered cell population. Protection for therapeutic cells and surviving cardiac cells can be achieved directly through mechanisms such as apoptosis reduction or indirectly through neovascularisation stimulation for example. Control of growth factor release can result from simple entrapment within the biomaterial or through more sophisticated approaches that exploit ionic or covalent attachment to regions of the scaffold.

A matrix metalloproteinase-sensitive PEG hydrogel was used to deliver the pro-angiogenic and pro-survival thymosin β4 in conjunction with hESC-derived vascular cells immediately after infarct induction in nude rats (Kraehenbuehl et al. 2011) (Table 10.3). Cell retention was not assessed but the combined treatment was determined by MRI to more effectively preserve EF and reduce diastolic dilation than the single treatments. Additionally greater neovascularisation was observed with evidence that at least a portion of the new vessels were derived from the delivered vascular cells.

BMMNC were delivered to fresh infarcts in mice using a PEGylated fibrin gel that had covalently incorporated HGF, a growth factor with multiple protective actions (Zhang et al. 2008). Histological analysis showed a 15-fold increase in engraftment at 4 weeks for the combination group relative to the 0.1 % seen in the cell alone group. EF was only significantly increased at 4 weeks and apoptosis reduced both remote to and within the infarct by the combination. PEGylated fibrin hydrogels were also utilised to inject engineered iPS into infarcted SCID mouse hearts (Bearzi et al. 2014). Here growth factors were not specifically released from the biomaterial but rather from the encapsulated cells that had been engineered to express MMP-9 and PLGF. The latter was introduced for its pro-angiogenic effect and the former to take advantage of its potential to break down the scar. Cell retention was not quantified but only when both proteins were secreted together was increases in vascularisation and decreases in fibrosis observed. All single treatments and the combined group equally increased fractional shortening (FS) above that of control but only the latter restored the velocity of blood flow in the left ventricular outflow to almost physiological levels.

The SAP RAD16-II nanofiber system has also been used to supply factors and cells together to infarcted hearts. In an early study, the cytoprotective IGF-1 was biotinylated and bound into biotinylated RAD16-II scaffold via a streptavidin sandwich (Davis et al. 2006). Binding IGF was found to moderately decrease apoptosis and increase engraftment of GFP-labelled neonatal cardiomyocytes 14 days after injection into rat hearts relative to untethered IGF or SAP alone. Only the tethered IGF was seen to improve FS 21 days after injection into infarcted rat hearts. The same system was used for introduction of CPC to infarcted rat hearts (Padin-Iruegas et al. 2009). Again the combination of IGF-1 and CPC was found to better preserve cardiac function and also increase vascularisation and reduce infarct size. The RAD16-II peptide was recently functionalised to incorporate a peptide mimic of the Notch1 ligand Jagged1 (Boopathy et al. 2014). Notch1 signaling is known to have a critical role in cardiac development and survival and differentiation of CPC. The modified SAP was injected with CPC during reperfusion in a rat model of ischemia/reperfusion. Importantly, global retention of CPC labelled with the fluorescent membrane label DiR was analysed in an in vivo

Table 10.3 Cell delivery within biomaterials containing growth factors for cardiac repair

Cell type	Model	Biomaterial	Factor	Duration	Retention	Efficacy	Reference
hESCDVC	Rat	MMP-sensitive PEG	Thymosin β4	6 weeks	ND	EF: Significant improvement for cells, growth factor and biomaterial combined compared to controls	Kraehenbuehl et al. (2011)
BMMNC	Murine	PEGylated fibrin gel	HGF	4 weeks	15-fold increase in engraftment (cells, growth factor and biomaterial)	EF: Significant improvement for cells, growth factor and biomaterial combined compared to controls	Zhang et al. (2008)
iPSC	Murine	PEG-fibrinogen scaffold	PIGF	30 days	ND	FS: 32.3 % (cells, growth factor and biomaterial), 25.1 % (Growth factor and biomaterial), 30.3 % (Cells)	Bearzi et al. (2014)
CMC	Rat	Self-assembling peptides	IGF-1	21 days	Growth factor improved cell retention by 25 %	FS: 45 % (cells, growth factor and biomaterial), 36 % (cells and biomaterial)	Davis et al. (2006)
CPC	Rat	Self-assembling peptides	IGF-1	4 weeks	2.4-fold increase in cell replication (cells, growth factor and biomaterial)	EF: Significant improvement for cells, growth factor and biomaterial combined compared to controls	Padin-Iruegas et al. (2009)
CPC	Rat	Self-assembling peptides	Notch1 ligand Jagged1 peptide	21 days	≈20 % (cells, peptide and biomaterial)	EF: Significant improvement for cells, peptide and biomaterial combined compared to controls	Boopathy et al. (2014)

hESCDVC human embryonic stem cell-derived vascular cells, *iPSC* induced pluripotent stem cells, *CMC* cardiac myocytes, *CPC* cardiac progenitor cells imaging system. Retention in the peptide mimic group relative to scaffolds either containing a scrambled version of the mimic or only RAD16-II was significantly elevated at above 80 % for the first 7 days but then fell to the 20 % levels observed in the others. Improved cardiac function as determined by haemodynamic parameters such as EF, stroke work and cardiac output was only observed in the peptide mimic/cell group. Similarly infarct size and vascularization were seen be reduced and increased respectively. In vitro studies suggested a paracrine effect as immobilized Jagged-1 mimic was shown to increase PDGF-BB and SCF expression by the CPC.

Survival and engraftment of delivered cells may well need utilization of a range of growth factors and this approach was explored in a study by Laflamme et al. (2007). In initial investigations, factors that influence a number of pathways such as anoikis, inflammation and ischemia were delivered in conjunction with hESC-derived cardiomyocytes to the hearts of athymic rats 4 days after they had undergone ischaemia-reperfusion. None of the treatments increased retention 1 week later as determined by in situ hybridization with a human-specific pan-centromeric genomic probe. This led to the development of a prosurvival cocktail that included Matrigel and a variety of other factors that simultaneously targeted several key points in cell death pathways. The cocktail resulted in a fourfold increase in engraftment relative to Matrigel and cells alone and this enhanced engraftment was associated with increased EF and wall thickness.

It is apparent from above that a variety of biomaterial formulations have enhanced both the engraftment of cells and their therapeutic efficacy. However, there are several aspects that now require optimisation after this initial period of investigation. In the majority of studies thus far where biomaterials have been used as cellular vehicles, engraftment has been assessed by histological based procedures. It would seem necessary to confirm these findings with the more quantitative global assays, ideally utilising in vivo imaging of cells stably expressing relevant reporter genes. Also at this point, very few large animal model studies have been performed but it might be expected that with the positive outcomes reported in small animals that this situation will now start to change. Finally, it may be prudent that direct comparative studies between the various optimal candidates are done before proceeding to the clinic. This may help avoid further disappointments in a field that is already somewhat beleaguered by setbacks.

References

Abbott JD, Huang Y, Liu D, Hickey R, Krause DS, Giordano FJ (2004) Stromal cell-derived factor-1alpha plays a critical role in stem cell recruitment to the heart after myocardial infarction but is not sufficient to induce homing in the absence of injury. Circulation 110(21):3300–3305

Azene N, Fu Y, Maurer J, Kraitchman DL (2014) Tracking of stem cells in vivo for cardiovascular applications. J Cardiovasc Magn Reson 16:7

Bastings MM, Koudstaal S, Kieltyka RE, Nakano Y, Pape AC, Feyen DA et al (2014) A fast pH-switchable and self-healing supramolecular hydrogel carrier for guided, local catheter injection in the infarcted myocardium. Adv Healthc Mater 3(1):70–78

Bearzi C, Gargioli C, Baci D, Fortunato O, Shapira-Schweitzer K, Kossover O et al (2014) PlGF-MMP9-engineered iPS cells supported on a PEG-fibrinogen hydrogel scaffold possess an enhanced capacity to repair damaged myocardium. Cell Death Dis 5

Bergmann O, Bhardwaj RD, Bernard S, Zdunek S, Barnabe-Heider F, Walsh S et al (2009) Evidence for cardiomyocyte renewal in humans. Science 324(5923):98–102

Bonios M, Terrovitis J, Chang CY, Engles JM, Higuchi T, Lautamaeki R et al (2011) Myocardial substrate and route of administration determine acute cardiac retention and lung bio-distribution of cardiosphere-derived cells. J Nucl Cardiol 18(3):443–450

Boopathy AV, Che PL, Somasuntharam I, Fiore VF, Cabigas EB, Ban K et al (2014) The modulation of cardiac progenitor cell function by hydrogel-dependent Notch1 activation. Biomaterials 35(28):8103–8112

Campbell NG, Suzuki K (2012) Cell Delivery routes for stem cell therapy to the heart: current and future approaches. J Cardiovasc Transl Res 5(5):713–726

Chang MY, Chang CH, Chen CH, Cheng B, Lin YD, Luo CY et al (2015) The time window for therapy with Peptide nanofibers combined with autologous bone marrow cells in pigs after acute myocardial infarction. PLoS One 10(3), e0115430

Chen CH, Wang SS, Wei EI, Chu TY, Hsieh PC (2013) Hyaluronan enhances bone marrow cell therapy for myocardial repair after infarction. Mol Ther 21(3):670–679

Chen IY, Wu JC (2013) Molecular imaging: the key to advancing cardiac stem cell therapy. Trends Cardiovasc Med 23(6):201–210

Cheng K, Blusztajn A, Shen D, Li T-S, Sun B, Galang G et al (2012) Functional performance of human cardiosphere-derived cells delivered in an in situ polymerizable hyaluronan-gelatin hydrogel. Biomaterials 33(21):5317–5324

Chenite A, Chaput C, Wang D, Combes C, Buschmann MD, Hoemann CD et al (2000) Novel injectable neutral solutions of chitosan form biodegradable gels in situ. Biomaterials 21(21):2155–2161

Christman KL, Fok HH, Sievers RE, Fang Q, Lee RJ (2004a) Fibrin glue alone and skeletal myoblasts in a fibrin scaffold preserve cardiac function after myocardial infarction. Tissue Eng 10(3–4):403–409

Christman KL, Vardanian AJ, Fang QZ, Sievers RE, Fok HH, Lee RJ (2004b) Injectable fibrin scaffold improves cell transplant survival, reduces infarct expansion, and induces neovasculature formation in ischemic myocardium. J Am Coll Cardiol 44(3):654–660

Clifford DM, Fisher SA, Brunskill SJ, Doree C, Mathur A, Watt S et al (2012) Stem cell treatment for acute myocardial infarction. Cochrane Database Syst Rev 2, CD006536

Cui XJ, Xie H, Wang HJ, Guo HD, Zhang JK, Wang C et al (2010) Transplantation of mesenchymal stem cells with self-assembling polypeptide scaffolds is conducive to treating myocardial infarction in rats. Tohoku J Exp Med 222(4):281–289

Dai W, Hale SL, Kay GL, Jyrala AJ, Kloner RA (2009) Delivering stem cells to the heart in a collagen matrix reduces relocation of cells to other organs as assessed by nanoparticle technology. Regen Med 4(3):387–395

Danoviz ME, Nakamuta JS, Marques FLN, dos Santos L, Alvarenga EC, dos Santos AA et al (2010) Rat adipose tissue-derived stem cells transplantation attenuates cardiac dysfunction post infarction and biopolymers enhance cell retention. PLoS One 5(8), e12077

Davis ME, Hsieh PC, Takahashi T, Song Q, Zhang S, Kamm RD et al (2006) Local myocardial insulin-like growth factor 1 (IGF-1) delivery with biotinylated peptide nanofibers improves cell therapy for myocardial infarction. Proc Natl Acad Sci U S A 103(21):8155–8160

de Jong R, Houtgraaf JH, Samiei S, Boersma E, Duckers HJ (2014) Intracoronary stem cell infusion after acute myocardial infarction: a meta-analysis and update on clinical trials. Circ Cardiovasc Interv 7(2):156–167

Dobner S, Bezuidenhout D, Govender P, Zilla P, Davies N (2009) A synthetic non-degradable polyethylene glycol hydrogel retards adverse post-infarct left ventricular remodeling. J Card Fail 15(7):629–636

Doyle B, Kemp BJ, Chareonthaitawee P, Reed C, Schmeckpeper J, Sorajja P et al (2007) Dynamic tracking during intracoronary injection of 18F-FDG-labeled progenitor cell therapy for acute myocardial infarction. J Nucl Med 48(10):1708–1714

Ehrbar M, Sala A, Lienemann P, Ranga A, Mosiewicz K, Bittermann A et al (2011) Elucidating the role of matrix stiffness in 3D cell migration and remodeling. Biophys J 100(2):284–293

Fisher SA, Doree C, Mathur A, Martin-Rendon E (2015) Meta-analysis of cell therapy trials for patients with heart failure. Circ Res 116(8):1361–1377

Frangogiannis NG (2006) The mechanistic basis of infarct healing. Antioxid Redox Signal 8(11–12):1907–1939

Frey N, Linke A, Suselbeck T, Muller-Ehmsen J, Vermeersch P, Schoors D et al (2014) Intracoronary delivery of injectable bioabsorbable scaffold (IK-5001) to treat left ventricular remodeling after ST-elevation myocardial infarction: a first-in-man study. Circ Cardiovasc Interv 7(6):806–812

Freyman T, Polin G, Osman H, Crary J, Lu M, Cheng L et al (2006) A quantitative, randomized study evaluating three methods of mesenchymal stem cell delivery following myocardial infarction. Eur Heart J 27(9):1114–1122

Furlani D, Ugurlucan M, Ong L, Bieback K, Pittermann E, Westien I et al (2009) Is the intravascular administration of mesenchymal stem cells safe? Mesenchymal stem cells and intravital microscopy. Microvasc Res 77(3):370–376

George JC, Goldberg J, Joseph M, Abdulhameed N, Crist J, Das H et al (2008) Transvenous intramyocardial cellular delivery increases retention in comparison to intracoronary delivery in a porcine model of acute myocardial infarction. J Interv Cardiol 21(5):424–431

Gnecchi M, Zhang Z, Ni A, Dzau VJ (2008) Paracrine mechanisms in adult stem cell signaling and therapy. Circ Res 103(11):1204–1219

Guo HD, Cui GH, Wang HJ, Tan YZ (2010) Transplantation of marrow-derived cardiac stem cells carried in designer self-assembling peptide nanofibers improves cardiac function after myocardial infarction. Biochem Biophys Res Commun 399(1):42–48

Gyongyosi M, Blanco J, Marian T, Tron L, Petnehazy O, Petrasi Z et al (2008) Serial noninvasive in vivo positron emission tomographic tracking of percutaneously intramyocardially injected autologous porcine mesenchymal stem cells modified for transgene reporter gene expression. Circ Cardiovasc Imaging 1(2):94–103

Gyongyosi M, Wojakowski W, Lemarchand P, Lunde K, Tendera M, Bartunek J et al (2015) Meta-analysis of Cell-based CaRdiac stUdiEs (ACCRUE) in patients with acute myocardial infarction based on individual patient data. Circ Res 116(8):1346–1360

Hale SL, Dai W, Dow JS, Kloner RA (2008) Mesenchymal stem cell administration at coronary artery reperfusion in the rat by two delivery routes: a quantitative assessment. Life Sci 83(13–14):511–515

Hofmann M, Wollert KC, Meyer GP, Menke A, Arseniev L, Hertenstein B et al (2005) Monitoring of bone marrow cell homing into the infarcted human myocardium. Circulation 111(17): 2198–2202

Holmes JW, Borg TK, Covell JW (2005) Structure and mechanics of healing myocardial infarcts. Annu Rev Biomed Eng 7:223–253

Hong KU, Guo Y, Li QH, Cao P, Al-Maqtari T, Vajravelu BN et al (2014a) c-kit+ Cardiac stem cells alleviate post-myocardial infarction left ventricular dysfunction despite poor engraftment and negligible retention in the recipient heart. PLoS One 9(5), e96725

Hong SJ, Hou D, Brinton TJ, Johnstone B, Feng D, Rogers P et al (2014b) Intracoronary and retrograde coronary venous myocardial delivery of adipose-derived stem cells in swine infarction lead to transient myocardial trapping with predominant pulmonary redistribution. Catheter Cardiovasc Interv 83(1):E17–E25

Hou DM, Youssef EAS, Brinton TJ, Zhang P, Rogers P, Price ET et al (2005) Radiolabeled cell distribution after intramyocardial, intracoronary, and interstitial retrograde coronary venous delivery—Implications for current clinical trials. Circulation 112(9):I150–I156

Ifkovits JL, Tous E, Minakawa M, Morita M, Robb JD, Koomalsingh KJ et al (2010) Injectable hydrogel properties influence infarct expansion and extent of postinfarction left ventricular remodeling in an ovine model. Proc Natl Acad Sci U S A 107(25):11507–11512

Jiang XJ, Wang T, Li XY, Wu DQ, Zheng ZB, Zhang JF et al (2009) Injection of a novel synthetic hydrogel preserves left ventricle function after myocardial infarction. J Biomed Mater Res A 90(2):472–477

Kadner K, Dobner S, Franz T, Bezuidenhout D, Sirry MS, Zilla P et al (2012) The beneficial effects of deferred delivery on the efficiency of hydrogel therapy post myocardial infarction. Biomaterials 33(7):2060–2066

Kang WJ, Kang HJ, Kim HS, Chung JK, Lee MC, Lee DS (2006) Tissue distribution of 18F-FDG-labeled peripheral hematopoietic stem cells after intracoronary administration in patients with myocardial infarction. J Nucl Med 47(8):1295–1301

Kharlamov AN, Duckers HJ, van Beusekom HM, Smits PC, Perin EC, Serruys PW (2013) Do we have a future with transcatheter adventitial delivery of stem cells? Int J Cardiol 165(2):217–221

Klein HM, Ghodsizad A, Marktanner R, Poll L, Voelkel T, Mohammad Hasani MR et al (2007) Intramyocardial implantation of CD133+ stem cells improved cardiac function without bypass surgery. Heart Surg Forum 10(1):E66–E69

Kraehenbuehl TP, Ferreira LS, Hayward AM, Nahrendorf M, van der Vlies AJ, Vasile E et al (2011) Human embryonic stem cell-derived microvascular grafts for cardiac tissue preservation after myocardial infarction. Biomaterials 32(4):1102–1109

Kraitchman DL, Tatsumi M, Gilson WD, Ishimori T, Kedziorek D, Walczak P et al (2005) Dynamic imaging of allogeneic mesenchymal stem cells trafficking to myocardial infarction. Circulation 112(10):1451–1461

Laflamme MA, Chen KY, Naumova AV, Muskheli V, Fugate JA, Dupras SK et al (2007) Cardiomyocytes derived from human embryonic stem cells in pro-survival factors enhance function of infarcted rat hearts. Nat Biotechnol 25(9):1015–1024

Landa N, Miller L, Feinberg MS, Holbova R, Shachar M, Freeman I et al (2008) Effect of injectable alginate implant on cardiac remodeling and function after recent and old infarcts in rat. Circulation 117(11):1388–1396

Lee LC, Wall ST, Klepach D, Ge L, Zhang Z, Lee RJ et al (2013) Algisyl-LVR with coronary artery bypass grafting reduces left ventricular wall stress and improves function in the failing human heart. Int J Cardiol 168(3):2022–2028

Lee RH, Pulin AA, Seo MJ, Kota DJ, Ylostalo J, Larson BL et al (2009) Intravenous hMSCs improve myocardial infarction in mice because cells embolized in lung are activated to secrete the anti-inflammatory protein TSG-6. Cell Stem Cell 5(1):54–63

Leor J, Tuvia S, Guetta V, Manczur F, Castel D, Willenz U et al (2009) Intracoronary injection of in situ forming alginate hydrogel reverses left ventricular remodeling after myocardial infarction in Swine. J Am Coll Cardiol 54(11):1014–1023

Lepperhof V, Polchynski O, Kruttwig K, Bruggemann C, Neef K, Drey F et al (2014) Bioluminescent imaging of genetically selected induced pluripotent stem cell-derived cardiomyocytes after transplantation into infarcted heart of syngeneic recipients. PLoS One 9(9), e107363

Li S-H, Lai TYY, Sun Z, Han M, Moriyama E, Wilson B et al (2009) Tracking cardiac engraftment and distribution of implanted bone marrow cells: comparing intra-aortic, intravenous, and intramyocardial delivery. J Thorac Cardiovasc Surg 137(5):1225–1233

Li X, Zhou J, Liu Z, Chen J, Lu S, Sun H et al (2014) A PNIPAAm-based thermosensitive hydrogel containing SWCNTs for stem cell transplantation in myocardial repair. Biomaterials 35(22): 5679–5688

Li Y, Yao Y, Sheng Z, Yang Y, Ma G (2011) Dual-modal tracking of transplanted mesenchymal stem cells after myocardial infarction. Int J Nanomedicine 6:815–823

Lin Y-D, Yeh M-L, Yang Y-J, Tsai D-C, Chu T-Y, Shih Y-Y et al (2010) Intramyocardial peptide nanofiber injection improves postinfarction ventricular remodeling and efficacy of bone marrow cell therapy in pigs. Circulation 122(11):S132–S141

Liu J, Narsinh KH, Lan F, Wang L, Nguyen PK, Hu S et al (2012a) Early stem cell engraftment predicts late cardiac functional recovery: preclinical insights from molecular imaging. Circ Cardiovasc Imaging 5(4):481–490

Liu Z, Wang H, Wang Y, Lin Q, Yao A, Cao F et al (2012b) The influence of chitosan hydrogel on stem cell engraftment, survival and homing in the ischemic myocardial microenvironment. Biomaterials 33(11):3093–3106

Lu S, Wang H, Lu W, Liu S, Lin Q, Li D et al (2010) Both the transplantation of somatic cell nuclear transfer- and fertilization-derived mouse embryonic stem cells with temperature-responsive chitosan hydrogel improve myocardial performance in infarcted rat hearts. Tissue Eng Part A 16(4):1303–1315

Lu WN, Lu SH, Wang HB, Li DX, Duan CM, Liu ZQ et al (2009) Functional improvement of infarcted heart by co-injection of embryonic stem cells with temperature-responsive chitosan hydrogel. Tissue Eng Part A 15(6):1437–1447

Luna SM, Silva SS, Gomes ME, Mano JF, Reis RL (2011) Cell adhesion and proliferation onto chitosan-based membranes treated by plasma surface modification. J Biomater Appl 26(1):101–116

Mitchell AJ, Sabondjian E, Sykes J, Deans L, Zhu W, Lu X et al (2010) Comparison of initial cell retention and clearance kinetics after subendocardial or subepicardial injections of endothelial progenitor cells in a canine myocardial infarction model. J Nucl Med 51(3):413–417

Muller-Ehmsen J, Krausgrill B, Burst V, Schenk K, Neisen UC, Fries JW et al (2006) Effective engraftment but poor mid-term persistence of mononuclear and mesenchymal bone marrow cells in acute and chronic rat myocardial infarction. J Mol Cell Cardiol 41(5):876–884

Murry CE, Soonpaa MH, Reinecke H, Nakajima H, Nakajima HO, Rubart M et al (2004) Haematopoietic stem cells do not transdifferentiate into cardiac myocytes in myocardial infarcts. Nature 428(6983):664–668

Nakamuta JS, Danoviz ME, Marques FLN, dos Santos L, Becker C, Goncalves GA et al (2009) Cell therapy attenuates cardiac dysfunction post myocardial infarction: effect of timing, routes of injection and a fibrin scaffold. PLoS One 4(6), e6005

Nelson DM, Ma Z, Fujimoto KL, Hashizume R, Wagner WR (2011) Intra-myocardial biomaterial injection therapy in the treatment of heart failure: materials, outcomes and challenges. Acta Biomater 7(1):1–15

Opie LH, Commerford PJ, Gersh BJ, Pfeffer MA (2006) Controversies in ventricular remodelling. Lancet 367(9507):356–367

Orlic D, Kajstura J, Chimenti S, Jakoniuk I, Anderson SM, Li B et al (2001) Bone marrow cells regenerate infarcted myocardium. Nature 410(6829):701–705

Padin-Iruegas ME, Misao Y, Davis ME, Segers VF, Esposito G, Tokunou T et al (2009) Cardiac progenitor cells and biotinylated insulin-like growth factor-1 nanofibers improve endogenous and exogenous myocardial regeneration after infarction. Circulation 120(10):876–887

Panda NC, Zuckerman ST, Mesubi OO, Rosenbaum DS, Penn MS, Donahue JK et al (2014) Improved conduction and increased cell retention in healed MI using mesenchymal stem cells suspended in alginate hydrogel. J Interv Card Electrophysiol 41(2):117–127

Pek YS, Wan AC, Ying JY (2010) The effect of matrix stiffness on mesenchymal stem cell differentiation in a 3D thixotropic gel. Biomaterials 31(3):385–391

Perin EC, Dohmann HF, Borojevic R, Silva SA, Sousa AL, Mesquita CT et al (2003) Transendocardial, autologous bone marrow cell transplantation for severe, chronic ischemic heart failure. Circulation 107(18):2294–2302

Perin EC, Tian M, Marini FC 3rd, Silva GV, Zheng Y, Baimbridge F et al (2011) Imaging long-term fate of intramyocardially implanted mesenchymal stem cells in a porcine myocardial infarction model. PLoS One 6(9), e22949

Pompilio G, Steinhoff G, Liebold A, Pesce M, Alamanni F, Capogrossi MC et al (2008) Direct minimally invasive intramyocardial injection of bone marrow-derived AC133+ stem cells in patients with refractory ischemia: preliminary results. Thorac Cardiovasc Surg 56(2):71–76

Radisic M, Christman KL (2013) Materials science and tissue engineering: repairing the heart. Mayo Clin Proc 88(8):884–898

Rane AA, Chuang JS, Shah A, Hu DP, Dalton ND, Gu Y et al (2011) Increased infarct wall thickness by a bio-inert material is insufficient to prevent negative left ventricular remodeling after myocardial infarction. PLoS One 6(6), e21571

Ransohoff KJ, Wu JC (2010) Advances in cardiovascular molecular imaging for tracking stem cell therapy. Thromb Haemost 104(1):13–22

Sanganalmath SK, Bolli R (2013) Cell therapy for heart failure: a comprehensive overview of experimental and clinical studies, current challenges, and future directions. Circ Res 113(6):810–834

Schachinger V, Aicher A, Dobert N, Rover R, Diener J, Fichtlscherer S et al (2008) Pilot trial on determinants of progenitor cell recruitment to the infarcted human myocardium. Circulation 118(14):1425–1432

Seif-Naraghi SB, Singelyn JM, Salvatore MA, Osborn KG, Wang JJ, Sampat U et al (2013) Safety and efficacy of an injectable extracellular matrix hydrogel for treating myocardial infarction. Sci Transl Med 5(173):173ra25

Senyo SE, Steinhauser ML, Pizzimenti CL, Yang VK, Cai L, Wang M et al (2013) Mammalian heart renewal by pre-existing cardiomyocytes. Nature 493(7432):433–436

Sheikh AY, Lin SA, Cao F, Cao Y, van der Bogt KE, Chu P et al (2007) Molecular imaging of bone marrow mononuclear cell homing and engraftment in ischemic myocardium. Stem Cells 25(10):2677–2684

Shen D, Cheng K, Marban E (2012) Dose-dependent functional benefit of human cardiosphere transplantation in mice with acute myocardial infarction. J Cell Mol Med 16(9):2112–2116

Silva SA, Sousa AL, Haddad AF, Azevedo JC, Soares VE, Peixoto CM et al (2009) Autologous bone-marrow mononuclear cell transplantation after acute myocardial infarction: comparison of two delivery techniques. Cell Transplant 18(3):343–352

Siminiak T, Fiszer D, Jerzykowska O, Grygielska B, Rozwadowska N, Kalmucki P et al (2005) Percutaneous trans-coronary-venous transplantation of autologous skeletal myoblasts in the treatment of post-infarction myocardial contractility impairment: the POZNAN trial. Eur Heart J 26(12):1188–1195

Singelyn JM, Sundaramurthy P, Johnson TD, Schup-Magoffin PJ, Hu DP, Faulk DM et al (2012) Catheter-deliverable hydrogel derived from decellularized ventricular extracellular matrix increases endogenous cardiomyocytes and preserves cardiac function post-myocardial infarction. J Am Coll Cardiol 59(8):751–763

Smith RR, Barile L, Cho HC, Leppo MK, Hare JM, Messina E et al (2007) Regenerative potential of cardiosphere-derived cells expanded from percutaneous endomyocardial biopsy specimens. Circulation 115(7):896–908

Strauer BE, Steinhoff G (2011) 10 years of intracoronary and intramyocardial bone marrow stem cell therapy of the heart: from the methodological origin to clinical practice. J Am Coll Cardiol 58(11):1095–1104

Sutton MG, Sharpe N (2000) Left ventricular remodeling after myocardial infarction: pathophysiology and therapy. Circulation 101(25):2981–2988

Swijnenburg R-J, Govaert JA, van der Bogt KEA, Pearl JI, Huang M, Stein W et al (2010) Timing of bone marrow cell delivery has minimal effects on cell viability and cardiac recovery after myocardial infarction. Circ Cardiovasc Imaging 3(1):77–85

Templin C, Zweigerdt R, Schwanke K, Olmer R, Ghadri JR, Emmert MY et al (2012) Transplantation and tracking of human-induced pluripotent stem cells in a pig model of myocardial infarction: assessment of cell survival, engraftment, and distribution by hybrid single photon emission computed tomography/computed tomography of sodium iodide symporter transgene expression. Circulation 126(4):430–439

Terrovitis J, Lautamaeki R, Bonios M, Fox J, Engles JM, Yu J et al (2009) Noninvasive quantification and optimization of acute cell retention by in vivo positron emission tomography after intramyocardial cardiac-derived stem cell delivery. J Am Coll Cardiol 54(17):1619–1626

Tossios P, Krausgrill B, Schmidt M, Fischer T, Halbach M, Fries JWU et al (2008) Role of balloon occlusion for mononuclear bone marrow cell deposition after intracoronary injection in pigs with reperfused myocardial infarction. Eur Heart J 29(15):1911–1921

Vulliet PR, Greeley M, Halloran SM, MacDonald KA, Kittleson MD (2004) Intra-coronary arterial injection of mesenchymal stromal cells and microinfarction in dogs. Lancet 363(9411):783–784

Wall ST, Walker JC, Healy KE, Ratcliffe MB, Guccione JM (2006) Theoretical impact of the injection of material into the myocardium: a finite element model simulation. Circulation 114(24):2627–2635

Wang H, Liu Z, Li D, Guo X, Kasper FK, Duan C et al (2012) Injectable biodegradable hydrogels for embryonic stem cell transplantation: improved cardiac remodelling and function of myocardial infarction. J Cell Mol Med 16(6):1310–1320

Wang T, Jiang XJ, Tang QZ, Li XY, Lin T, Wu DQ et al (2009) Bone marrow stem cells implantation with alpha-cyclodextrin/MPEG-PCL-MPEG hydrogel improves cardiac function after myocardial infarction. Acta Biomater 5(8):2939–2944

Wenk JF, Wall ST, Peterson RC, Helgerson SL, Sabbah HN, Burger M et al (2009) A method for automatically optimizing medical devices for treating heart failure: designing polymeric injection patterns. J Biomech Eng 131(12):121011

Westrich J, Yaeger P, He C, Stewart J, Chen R, Seleznik G et al (2010) Factors affecting residence time of mesenchymal stromal cells (MSC) injected into the myocardium. Cell Transplant 19(8):937–948

Yang JJ, Liu ZQ, Zhang JM, Wang HB, Hu SY, Liu JF et al (2013) Real-time tracking of adipose tissue-derived stem cells with injectable scaffolds in the infarcted heart. Heart Vessels 28(3):385–396

Zhang G, Hu Q, Braunlin EA, Suggs LJ, Zhang J (2008) Enhancing efficacy of stem cell transplantation to the heart with a PEGylated fibrin biomatrix. Tissue Eng Part A 14(6):1025–1036

Chapter 11
Stem Cell Therapies for Cardiac Regeneration: Current Burden—Future Directions

Rosalinda Madonna

11.1 Stem Cell Transplantation: past and Present from the Clinical Trials

Despite contemporary medical treatments, heart failure remains a major cause of morbidity and mortality in developed countries (Go et al. 2014). Heart transplantation is a therapeutic option for patients at end-stage heart failure. However, because of the limited availability of donor organs, immune rejections, and infectious complications, alternative treatments are currently under evaluation. Transplantation of stem/progenitor cells have been considered alternative treatment for heart repair, attracting tremendous attention of basic scientists and clinicians (Forrester et al. 2003; Zimmermann et al. 2006; Sharma and Raghubir 2007). There are a number of clinical studies which have looked at cardiac function recovery via heart repair and regeneration by stem cell therapy in various clinical scenarios (Assmus et al. 2010; Bolli et al. 2011; Schaefer et al. 2011; Chugh et al. 2012; Makkar et al. 2012). The first report about stem cell therapy in heart failure utilized skeletal myoblasts in 1998 (Taylor et al. 1998). Later studies utilized total bone marrow (BM) stem cells (Perin et al. 2003; Willerson et al. 2010; Traverse et al. 2012, 2013), BM-derived mesenchymal stem cells (BM-MSCs), endothelial progenitor cells (EPCs), CD34+ and CD133+ cells (Sanganalmath and Bolli 2013). In addition, the potential of embryonic stem cells (van Laake et al. 2009), adipose tissue-derived MSCs (Perin et al. 2014), hematopoietic stem cells and cardiac stem (CSC) and progenitor cells (CPCs) (Smits et al. 2009) have been studied (Sanganalmath and Bolli 2013).

R. Madonna (✉)
Institute of Cardiology, Department of Neurosciences and Imaging Laboratory of Experimental Cardiology, Center of Excellence on Aging—CESI, "G. d'Annunzio" University, Chieti Scalo, Italy
e-mail: rmadonna@unich.it

R. Madonna (ed.), *Stem Cells and Cardiac Regeneration*, Stem Cell Biology and Regenerative Medicine, DOI 10.1007/978-3-319-25427-2_11

The REPAIR-AMI trial suggested that intracoronary delivery of BM cells in patients with acute myocardial infarction (AMI) resulted in significant improvements in cardiac function, which were preserved over 2 years (Assmus et al. 2010). In contrast, the BOOST trial, which was similarly constructed, showed based on 18-month and 5-year follow ups only an initial improvement with little sustained effect (Schaefer et al. 2011). The recent trials SCIPIO (Bolli et al. 2011; Chugh et al. 2012) and CADUCEUS (Makkar et al. 2012), showed promising results following the application of CSCs. Taken together, it appears that stem cells do contribute to cardiac repair or survival. However, it seems that the mechanism of action to date has not been differentiation into cardiomyocytes but indirect effects through the secretion of growth factors, at least for the cell types that are being used in clinical trials. Evidence is accumulating that these released factors direct a number of restorative processes including myocardial protection, neovascularization, and cardiac remodeling (Mirotsou et al. 2011). Novel delivery systems for enhancing the low stem cell engraftment and the importance of stem cell-mediated paracrine effects are discussed in Chaps. 13 and 14, respectively. With respect to stem cell therapy, studies that applied CPCs or preparations of exogenous stem cells originating from BM or peripheral blood, both positive and negative studies suffer from small cohort sizes, lack of long-term clinical outcome as endpoint and mechanism for the beneficial effect of exogenously applied stem cells. Nevertheless, most, but not all, studies demonstrated reduced infarct size and recovery of heart function (reviewed in: (Sanganalmath and Bolli 2013)). Several ongoing multi-center trials will provide more insight into the clinical applicability of stem cell therapy and may shed more light on the mechanism of this strategy, *e.g.* the involvement of paracrine effects rather than differentiation. There are a vast number of novel biological pathways targeted by secretome through a variety of SRM (*e.g.* proteins, microRNA, growth factors, antioxidants, proteasomes and exosomes) and novel options for advanced therapies (ex-vivo cell-based gene therapy for stem cell rejuvenation, biomaterials, etc.), reviewed above that have reached the phase of pre-clinical application and could be ready for clinical translation, after several questions will be answered, including: (1) what are the best methods to maximize the effects of enhanced stem cell therapy in vivo and are new technologies required to achieve this?; (2) how do the properties of the stem cell secretome (composition and sustainability) change in vitro and following transplantation? and (3) how do the secretome or genetically-modified stem cells or stem cells in conjunction with biomaterials influence the function of the local microenvironment and resident stem cells post-transplantation, and how can we optimize these local effects? In summary, traditional stem cell therapy as well as advanced stem cell therapy preparations are promising options for cardiac regeneration and repair and state-of-art reached the clinical translation phase, however, results are still controversial for traditional stem cell therapy and new protocols still wait to be set-up the new stem cell preparations. The reasons for some negative studies of traditional stem cell therapy can be attributed to the presence of different limitations such as mechanism(s) of action of stem cells, inadequate recruitment of circulating or resident cardiac stem cells; poor capability of adult stem cells to differentiate into cardiomyocytes; elevated mortality of transplanted stem cells; age-related changes, including increased rates of apoptosis and senescence; anomalous electro-mechanical behavior of transplanted cells, in addition to optimal cell type(s), and dose, route, and frequency of cell administration (reviewed in Chap. 10).

11.2 Stem Cell Aging and Rejuvenation

It is now known that stem cells are not exempt from aging (Lopez-Otin et al. 2014). As a result, resident stem/progenitor cells in elderly humans may have a decreased capacity for repair in response to tissue injury (Issa 2014; Lopez-Otin et al. 2014; Rando and Wyss-Coray 2014). Nevertheless, it has become increasingly clear that it is possible to reverse stem cell aging by rejuvenating existing aged cells (Rando and Chang 2012). This has been possible because of better understanding of the genes and signaling involved in stem cell aging. Pim-1 kinase has been identified as an anti-senescence and anti-apoptotic factor in CSCs and MSCs (Borillo et al. 2010; Mohsin et al. 2012, 2013). Genetic modification of aged human CPCs with Pim-1 kinase results in remarkable rejuvenation of the CPCs associated with enhanced proliferation, increased telomere lengths, and decreased susceptibility to replicative senescence (Mohsin et al. 2012, 2013). Notch plays also important roles in cardiac differentiation, regeneration and expansion of CPCs in mice (Yang et al. 2012; Nemir et al. 2014; Zhao et al. 2014) and zebrafish (Zhao et al. 2014). Activation of Notch restored "youthful" myogenic responses to satellite muscle cells isolated from 70-year-old humans, rendering them similar to cells from 20-year-old humans (Conboy and Rando 2012). The activation of telomere-telomerase axis contributes to cell survival and proliferation, and to prevent cellular senescence (Jan et al. 2011; Qu et al. 2011). A subpopulation of adipose tissue-derived MSCs (AT-MSCs) was recently identified that expresses high levels of myocardin (MYOCD), a nuclear transcription co-factor for myogenic and anti-apoptotic genes, and the catalytic subunit of telomerase (i.e., telomerase reverse transcriptase or TERT) (Madonna et al. 2012). AT-MSCs that co-express TERT and MYOCD have increased endogenous levels of octamer-binding transcription factor 4 (Oct-4), MYOCD, myocyte-specific enhancer factor 2c (Mef2c), and homeobox protein Nkx2.5 (Madonna et al. 2012), high cardiovascular regenerative potential (Madonna et al. 2013), as well as decreased frequencies of both spontaneous cell death and Fas-induced apoptosis (Madonna et al. 2013). The delivery of the TERT and MYOCD genes into AT-MSCs can restore MSCs from aged mice by increasing cell survival, proliferation, and smooth muscle myogenic differentiation in vitro (Madonna et al. 2013). Furthermore, the therapeutic efficacy of these rejuvenated cells was demonstrated in an in vivo hindlimb ischemia model (Madonna et al. 2013).

11.3 Current Burden and Future Directions

The following limitations of current approaches have been identified as yet to be solved: (1) inadequate recruitment of circulating or resident cardiac stem cells; (2) poor capability of adult stem cells to differentiate into cardiomyocytes; (3) elevated mortality of transplanted stem cells; (4) anomalous electro-mechanical behavior of transplanted cells after stimulation and the eventual onset of arrhythmias; (5) formation of new heart tissue structure differing from that of normal heart; and (6) diminished function of both resident and circulating stem/progenitor cells or even induced pluripotent stem cells with the onset of aging and age-related cardiovascular disease (Wang et al. 2011; Wu et al. 2011; Madonna et al. 2013; Pavo et al. 2014; Rohani et al. 2014). The question arose as to how stem cell therapy nevertheless can lead to recovery of cardiac function after ischemic injury. Attempts to answer this question have been made and rely on four possible different ways of using this strategy: **1.** inject stem-derived cellular products—the so called stem cell therapy without the cells—which consists in the use of the collected types of molecules released by the stem cells, called the secretome or stem cell released molecules (SRM), including proteins, microRNA, growth factors, antioxidants, proteasomes and exosomes targeting a multitude of biological pathways through paracrine actions of the transplanted or activated stem cells; **2.** boost the endogenous regenerative capacity of the adult heart, which retains some capacity of self-healing and self-renewal; **3.** transplant genetically modified stem cells, in which exogenous genes have been previously introduced by viral or not viral delivery vector, to exhibit delayed senescence, resistance to apoptosis and enhanced regenerative properties; **4.** transplant stem cells in conjunction with biocompatible injectable biomaterials that are capable to enhance stem cell retention and survival through a variety of mechanisms. Large clinical studies will be necessary to get more insight into the clinical applicability of these novel stem cell strategies, which would be expected to offer further opportunities to treatment of human patients with heart failure.

References

Assmus B, Rolf A et al (2010) Clinical outcome 2 years after intracoronary administration of bone marrow-derived progenitor cells in acute myocardial infarction. Circ Heart Fail 3(1):89–96

Bolli R, Chugh AR et al (2011) Cardiac stem cells in patients with ischaemic cardiomyopathy (SCIPIO): initial results of a randomised phase 1 trial. Lancet 378(9806):1847–1857

Borillo GA, Mason M et al (2010) Pim-1 kinase protects mitochondrial integrity in cardiomyocytes. Circ Res 106(7):1265–1274

Chugh AR, Beache GM et al (2012) Administration of cardiac stem cells in patients with ischemic cardiomyopathy: the SCIPIO trial: surgical aspects and interim analysis of myocardial function and viability by magnetic resonance. Circulation 126(11 Suppl 1):S54–S64

Conboy IM, Rando TA (2012) Heterochronic parabiosis for the study of the effects of aging on stem cells and their niches. Cell Cycle 11(12):2260–2267

Forrester JS, Price MJ et al (2003) Stem cell repair of infarcted myocardium: an overview for clinicians. Circulation 108(9):1139–1145
Go AS, Mozaffarian D et al (2014) Heart disease and stroke statistics—2014 update: a report from the American Heart Association. Circulation 129(3):e28–e292
Issa JP (2014) Aging and epigenetic drift: a vicious cycle. J Clin Invest 124(1):24–29
Jan HM, Wei MF et al (2011) The use of polyethylenimine-DNA to topically deliver hTERT to promote hair growth. Gene Ther 19(1):86–93
Lopez-Otin C, Blasco MA et al (2014) The hallmarks of aging. Cell 153(6):1194–1217
Madonna R, Wu D et al (2012) Myocardin-A enhances expression of promyogenic genes without depressing telomerase activity in adipose tissue-derived mesenchymal stem cells. Int J Cardiol 167(6):2912–2921
Madonna R, Taylor DA et al (2013) Transplantation of mesenchymal cells rejuvenated by the overexpression of telomerase and myocardin promotes revascularization and tissue repair in a murine model of hindlimb ischemia. Circ Res 113(7):902–914
Makkar RR, Smith RR et al (2012) Intracoronary cardiosphere-derived cells for heart regeneration after myocardial infarction (CADUCEUS): a prospective, randomised phase 1 trial. Lancet 379(9819):895–904
Mirotsou M, Jayawardena TM et al (2011) Paracrine mechanisms of stem cell reparative and regenerative actions in the heart. J Mol Cell Cardiol 50(2):280–289
Mohsin S, Khan M et al (2012) Human cardiac progenitor cells engineered with Pim-I kinase enhance myocardial repair. J Am Coll Cardiol 60(14):1278–1287
Mohsin S, Khan M et al (2013) Rejuvenation of human cardiac progenitor cells with Pim-1 kinase. Circ Res 113(10):1169–1179
Nemir M, Metrich M et al (2014) The Notch pathway controls fibrotic and regenerative repair in the adult heart. Eur Heart J 35(32):2174–2185
Pavo N, Charwat S et al (2014) Cell therapy for human ischemic heart diseases: critical review and summary of the clinical experiences. J Mol Cell Cardiol 75C:12–24
Perin EC, Dohmann HF et al (2003) Transendocardial, autologous bone marrow cell transplantation for severe, chronic ischemic heart failure. Circulation 107(18):2294–2302
Perin EC, Sanz-Ruiz R et al (2014) Adipose-derived regenerative cells in patients with ischemic cardiomyopathy: The PRECISE Trial. Am Heart J 168(1):88–95, e2
Qu Y, Duan Z et al (2011) Telomerase reverse transcriptase upregulation attenuates astrocyte proliferation and promotes neuronal survival in the hypoxic-ischemic rat brain. Stroke 42(12):3542–3550
Rando TA, Chang HY (2012) Aging, rejuvenation, and epigenetic reprogramming: resetting the aging clock. Cell 148(1–2):46–57
Rando TA, Wyss-Coray T (2014) Stem cells as vehicles for youthful regeneration of aged tissues. J Gerontol A Biol Sci Med Sci 69(Suppl 1):S39–S42
Rohani L, Johnson AA et al (2014) The aging signature: a hallmark of induced pluripotent stem cells? Aging Cell 13(1):2–7
Sanganalmath SK, Bolli R (2013) Cell therapy for heart failure: a comprehensive overview of experimental and clinical studies, current challenges, and future directions. Circ Res 113(6):810–834
Schaefer A, Zwadlo C et al (2011) Long-term effects of intracoronary bone marrow cell transfer on diastolic function in patients after acute myocardial infarction: 5-year results from the randomized-controlled BOOST trial—an echocardiographic study. Eur J Echocardiogr 11(2):165–171
Sharma R, Raghubir R (2007) Stem cell therapy: a hope for dying hearts. Stem Cells Dev 16(4):517–536
Smits AM, van Laake LW et al (2009) Human cardiomyocyte progenitor cell transplantation preserves long-term function of the infarcted mouse myocardium. Cardiovasc Res 83(3):527–535
Taylor DA, Atkins BZ et al (1998) Regenerating functional myocardium: improved performance after skeletal myoblast transplantation. Nat Med 4:929–933

Traverse JH, Henry TD et al (2012) Effect of the use and timing of bone marrow mononuclear cell delivery on left ventricular function after acute myocardial infarction: the TIME randomized trial. JAMA 308(22):2380–2389
Traverse JH, Henry TD et al (2013) One-year follow-up of intracoronary stem cell delivery on left ventricular function following ST-elevation myocardial infarction. JAMA 311(3):301–302
van Laake LW, Passier R et al (2009) Improvement of mouse cardiac function by hESC-derived cardiomyocytes correlates with vascularity but not graft size. Stem Cell Res 3(2–3):106–112
Wang B, Miyagoe-Suzuki Y et al (2011) Reprogramming efficiency and quality of induced Pluripotent Stem Cells (iPSCs) generated from muscle-derived fibroblasts of mdx mice at different ages. PLoS Curr 3, RRN1274
Willerson JT, Perin EC et al (2010) Intramyocardial injection of autologous bone marrow mononuclear cells for patients with chronic ischemic heart disease and left ventricular dysfunction (First Mononuclear Cells injected in the US [FOCUS]): rationale and design. Am Heart J 160(2):215–223
Wu KH, Mo XM et al (2011) Stem cell engraftment and survival in the ischemic heart. Ann Thorac Surg 92(5):1917–1925
Yang J, Bucker S et al (2012) Inhibition of Notch2 by Numb/Numblike controls myocardial compaction in the heart. Cardiovasc Res 96(2):276–285
Zhao L, Borikova AL et al (2014) Notch signaling regulates cardiomyocyte proliferation during zebrafish heart regeneration. Proc Natl Acad Sci U S A 111(4):1403–1408
Zimmermann WH, Didie M et al (2006) Heart muscle engineering: an update on cardiac muscle replacement therapy. Cardiovasc Res 71(3):419–429

Index

R. Madonna (ed.), *Stem Cells and Cardiac Regeneration*, Stem Cell Biology and Regenerative Medicine, DOI 10.1007/978-3-319-25427-2

Lightning Source UK Ltd.
Milton Keynes UK
UKOW05n1854211016

285873UK00010B/114/P

9 783319 254258